C0-AWC-956

So Much Data
So Little Math

How to Predict Data Trends
5 Easy Profitable Methods

William D. May

University Press of America, ® Inc.
Lanham • New York • Oxford

University Press of America, ® Inc.
4720 Boston Way
Lanham, Maryland 20706

12 Hid's Copse Rd.
Cumnor Hill, Oxford OX2 9JJ

Printed in the United States of America
British Library Cataloging in Publication Information Available

Library of Congress Cataloging-in-Publication Data

May, William D.
So much data so little math : how to predict data trends, 5 easy profitable methods / William D. May.
p. cm.
Includes bibliographical references and index.
l. Mathematical statistics. I. Title.
QA276.M375 2000 658.4'033—dc21 99-462020 CIP

ISBN 0-7618-1640-2 (pbk: alk. ppr.)

™ The paper used in this publication meets the minimum requirements of American National Standard for Information Sciences—Permanence of Paper for Printed Library Materials, ANSI Z39.48—1984

To the boys:

Chad, Clint, Cameron and Ross

Contents

Acknowledgements

The clipart in the figures is from the *ClickArt 200,000* graphics package by the Learning Company. The "Krista" graphic was assembled with the *PrintMaster Gold 4.0* software package.

I'd like to thank my wife Rhonda. Being married to a writer is hard!

Finally, the views expressed in this book are my own and do not necessarily reflect those of the United States Government or Virginia Tech.

Preface

Data is everywhere. There are profitable trends and patterns in that data. Yet if you never took the right math courses—what can you do about it? The Information Age is upon us. Today, many employees in large corporations, investors and small business owners are stymied because they can't predict trends in their data. Often, they blame their weak math. Until recently, they were right. However, recent breakthroughs in user-friendly software and new methods have placed trend analysis and predictions within the grasp of these people.

This book's program is for people that need *results now*, for people that want to try out mathematical modeling *before they commit to years* of formal math training. This book focuses on a program that will quickly give you the capability to analyze the data that you understand best and develop business models from it. A goal of the book is to give the reader a capability to develop a working mathematical model of their own business or investment environment.

This idea of "quick data analysis" with "little math" might seem like "something for nothing." It could sound a little suspicious; often how-to books are full of empty promises. However, it's not "something for nothing." There is a *tradeoff*. The tradeoff is *specific business area knowledge instead of advanced math*. Here is how it works: mathematicians and statisticians are generalists; you are a specialist in some business or investment area (I am assuming). Generalists—mathematicians and statisticians—choose data with some pre-modeling phase mathematics and they validate models with many post-modeling phase math techniques and statistical measures. Specialists can use their knowledge and business savvy to choose appropriate data in the pre-modeling phase. Specialists can also compare the model with reality in the post-modeling phase, without advanced mathematics. The modeling phase itself is normally done with computers that can be run without a detailed knowledge of calculus or statistics. If you really understand your area then you can do data modeling in that area with much less math than you might believe.

Suppose you are an adult whose math skills are very rusty or you are an adult who quit math ten years ago after algebra II. Suppose that now you want to complete math through college statistics—you have a formidable task. You need to review algebra then enroll and pass courses in trigonometry, precalculus, calculus and statistics. This could mean 4 to 6 courses spanning 2 to 3 years. That's a long time to wait if you want to start analyzing data for your own business or investments. You need an alternative—a program that is much quicker. The book's program offers an alternative path relying more on modern software than math skills. (You need some math, of course, to understand models and run the software. Overall, the program requires algebra II.)

There are real world case studies and examples throughout this book. It is here that most of the learning comes from. These examples will give you ideas about how to view your own data. They will give you ideas about which software techniques to use. They will give you ideas about what you can expect to get from data analysis. Your own data and needs are unique; you will have to come up with ideas on how to analyze them. It's the ideas in the examples that count most. These ideas will get you thinking about your unique problems. Author John Steinbeck put it best many years ago: "Ideas are like rabbits. You get a couple and learn how to handle them, and pretty soon you have a dozen."

WDM

So Much Data

So Little Math

How To Predict Data Trends

5 Easy Profitable Methods

Chapter 1

Data Analysis

What Is It? How Hard Is It? How Much Is It Worth?

Q I am in a hurry. You have 25 words or less to tell me what this data analysis book can do for me.

A You can quickly turn your data into trend predictions and profitable plans. You will be a better employee, investor or small business owner.

Q OK, but I still have a short attention span. Tell me in 25 words or less how this book can do all this.

A You will learn to make business models with software for prediction and planning. The book uses real-world examples, checklists and very little math.

Q Not bad. But, what I really want is graduate-degree statistician or MBA-level data analysis abilities. With *absolutely no* math. And I want to be done in a *few hours* during one weekend. Can you do this for me? I must warn you I am just browsing in a bookstore, I haven't bought this book yet, I can buy other books.

A Sorry I can't, no one can. It would be lying to say otherwise. No hard feelings though, let me recommend a book for you on science fiction and fantasy . . .

Q Just kidding! Maybe I could use this program. But I am worried about the word "math," I was good at math in high school but I don't want to read a math book now.

A This is *definitely not a math book.* Actually, a main goal is to allow the reader to develop good modeling capability without calculus and statistics. It turns out that modern software can help you do this. A reader can learn a few modeling techniques and how to apply them in the few hours it takes to read this book. Then the reader can decide which example and technique best fits their application. Next the reader can start analyzing data with their own software or go out and buy a new software package to do it. Results are quick. The book's examples, techniques and checklists are very general. After reading this book, you will be able to analyze data at a level you might think only mathematicians, statisticians and MBAs could attain.

Q This idea of "quick data analysis" with "little math" sounds to me like "something for nothing." I am a little suspicious. I hear many "get rich quick" schemes and often how-to books are full of empty promises. How do I know this book is different?

A Great question! It's not "something for nothing." There is a *tradeoff.* The tradeoff is *specific business area knowledge instead of advanced math.* Here is how it works: mathematicians and statisticians are generalists; you are a specialist in some business or investment area (I am assuming). Generalists—mathematicians and statisticians—choose data with some pre-modeling phase mathematics and they validate models with many post-modeling phase math techniques and statistical measures. Specialists can use their knowledge and business savvy to choose appropriate data in the pre-modeling phase. Specialists can also compare the model with reality in the post-modeling phase, without advanced mathematics. The modeling phase itself is normally done with computers that can be run without a detailed knowledge of calculus or statistics. If you really understand your area then you can do data modeling in that area with much less math than you might believe.

Data is everywhere. There are profitable trends and patterns in that data. Yet if you never took the right math courses—what can you do about it? The Information Age is upon us. Today, many employees in large corporations, investors and small business owners are stymied because they can't predict trends in their data. Often, they blame their weak math. Until recently, they were right. However, recent breakthroughs in user-friendly software and new methods have placed trend analysis and predictions within the grasp of these people. If you are one of these people then this book will show you how. This book's program is for people that need *results now*, for people that want to try out mathematical modeling *before they commit to years* of formal math training. This book focuses on a program that will quickly give you the capability to analyze the data that you understand best and develop business models from it. A goal of the book is to give the reader a capability to develop a working mathematical model of their own business or investment environment. We are in the information age; data is everywhere. Computer calculations, algorithms and models based on the information can help us solve business problems. There is a lot of money made each day just from using data analysis algorithms and models. Computer expert Robert Cross once summed it up nicely in USA Today: "It's like taking raw information and spinning money out of it."

Exactly what is a model? Well there isn't any all-inclusive definition but they have several characteristics. From an article in PC AI magazine: "Simulation, a key modeling tool, is an act of imitation." ... "For example, the board game Monopoly ™ is a model of a real system: the hotels and facilities of Atlantic City." Models are simulations of reality; they operate, react, progress and decline in ways that mimic that part of the real world that they are simulating. Models allow you to predict, to study trends, and to do tradeoff studies. This book is for knowledgeable but nonmathematical readers. It's for people that understand high school math and perhaps calculus or beginning statistics. It's for those who never understood how to apply these courses to all the data that surrounds them. It's for the self-employed who must analyze their own data and for the investor who wants to predict future trends. It's for almost everyone.

How You Use Models

A model of demand for donuts at a mall donut shop might include tomorrow's predicted temperature, rainfall, the day of the week and the shopping season. All these factors would be inputs or independent variables or predictor variables (terminology differs) to the model. They would be factors that effect the number of mall customers and their individual demand for donuts. These factors would determine the demand for donuts. This model could be a mathematical formula or computer program that predicts tomorrow's demand. This model would be critical to the success of the bakery: a bakery can't waste too many ingredients by over baking or risk alienating too many customers by under baking. Other examples would be a model of the potential profitability of a new store with independent variables like: density of population in a five-mile radius, number of competing stores in a five-mile radius and average income in that five-mile radius area. Or perhaps a model of the change in the corporate profit margin as a function of the yearly amount of employee technical retraining.

There are many uses of models; the most common are: trend prediction—making a model with past data and using it to predict future trends. Optimal policy decisions—making a model based on available data. Then using this model to try several different policies/strategies and determining which one gives the best results (e.g., optimal, most profit, least cost, fewest, maximum, minimum, etc.). Cost-benefit studies—make a model then examine different tradeoffs by varying the values of the input parameters.

The book's program starts with your business knowledge being used to wisely choose input data sets. Then these input data sets are loaded into a computer program of one of several types, regression, spreadsheet, database or neural network. Finally, the program is run with your data and the output is a model that you can use. Examples in following chapters will discuss three parts of the problem. First, examples will show the nature and form of the input data. Then different modeling software will be discussed. Finally, we will see different forms of your model (e.g., math function or computer program or set of rules). The following diagram summarizes the book's program. In the two step process of turning data into a plan we will look at how to choose and use data, then how to choose and use a modeling method. After we develop a model, we will see several examples of what you can profitably do with the model.

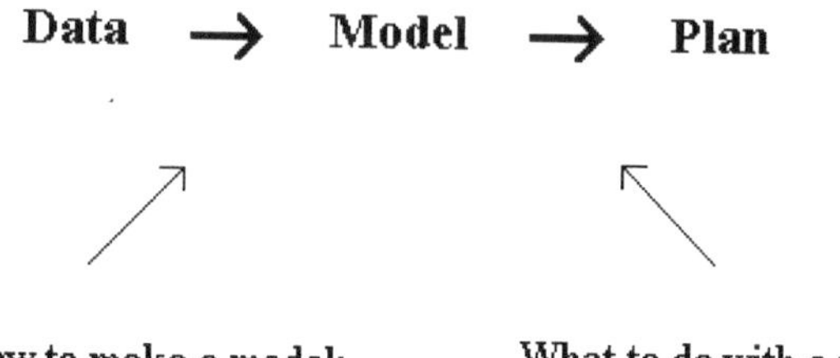

How to make a model:
- **choose input data**
- **choose method**
- **use software**

What to do with a model:
- **prediction**
- **trend analysis**
- **optimal plans**
- **cost-benefit tradeoffs**

Business Area Knowledge Instead Of Advanced Math

You have an advantage over a mathematician—a generalist. This is because you understand the processes and limits of your specific business area. This is also because you understand the data: which data effects other data and which data form an independent set of inputs. Finally, because you understand what the model should do even before you develop it. A mathematician needs calculus and statistics to understand and apply validation checks. You, the specialist, understand what is valid, but in a non-mathematical way. Because of all this, usually (caveat: not always) you can use software to develop a math model. Then you can do the pre-modeling data reduction and post-modeling validation with business practice understanding instead of advanced math. That means that you can develop good math models without having an advanced math background—modern software will be enough.

Show Me The Money

The more math you take the more income you will earn in a lifetime. It is the same for data analysis. That's the truth. Actually, much of the advantage more math confers on a person is an increased data analysis capability. Most of the general statistics are based on more math and technical training. Since that is close and since there aren't any good statistics based on just increased data analysis capabilities—let's look at some statistics. The Department of Labor released a study that showed that people with a solid high school math background (rigorous courses beyond algebra and geometry) had average lifetime earnings of 38% above that of others. Also in 1997 starting salaries for college graduates in Chemical Engineering were $46,000, for Physics, $36,000, for General Business Administration, $30,000 and for Retailing, $26,000. See the pattern. Data analysis is an important part of the advantage that more math gives you. However, for data analysis ability there are alternative ways to achieve it besides just taking more mathematics training. General statistics on how much more you make by knowing data analysis and prediction techniques are hard to come by. However, practical people want to see concrete examples anyway. We will see case studies throughout the rest of the book. These case studies will compare normal business judgement, simple data analysis methods and more complex data analysis. The figure of comparison will be dollars.

To be as up front as possible, some of these comparisons are listed in the following table. Now, we haven't yet considered these examples with their simulated business situations and data analysis methods. So for now, we will just present results. For now, we will leave unanswered questions about what analysis methods are being compared. Normal business judgement without any data modeling is the least profitable method in all these scenarios. Notice that data analysis doesn't rule out using business judgement too. As we will see, the people in our case studies applied both. Data analysis strengthens business judgement. *Data analysis does not replace* the need for *business judgement.* Look at the results. Often data analysis combined with business judgement can greatly outperform business judgement without any data analysis.

Normal Business Judgement

vs.

Book's Data Analysis Method

	Normal Business Judgement	Data Analysis Method
Bakery Scenario	$29,000	$84,000
Mail Order Scenario	$16,000	$61,000
Large Corporation Scenario	loss of hundreds of thousands	$7000 profit per employee

Later we will see these tables in full. There are many other levels of data analysis that will also be compared. For now notice that business judgement plus data analysis has profit margins of 290% and 380% higher than without data analysis. In addition, in the large corporate example, the difference is between a profit and a loss. These are simulations of real world situations. These examples are like your data problems. Most important, these data analysis methods are entirely within your grasp.

Modeling with data analysis is profitable—that's the main point. These results support that. Normal business judgements based on rules-of-thumb and high level concepts often fail in comparison to simple but solid data models. And that's the bottom line.

What About The Math?

Math is always an issue. People don't like it or feel that math is too hard and most people don't feel that they have had enough. Most educators and business leaders agree—the American working force, as a whole, lacks the necessary math skills. Therefore, maybe you believe, that millions of people should rush out and enroll in trigonometry, calculus and statistics classes. Now for reality, math education ages fast. People quickly forget basic algebra and geometry techniques. If you have been out of school for ten years you can't just jump back in and take the next math course in the math sequence. You must first review old math classes. Suppose you are an adult whose math skills are very rusty or you are an adult who quit math ten years ago after algebra II. Suppose that now you want to complete math through college statistics—you have a formidable task. You need to review algebra then enroll and pass courses in trigonometry, pre-calculus, calculus and statistics. This could mean 4 to 6 courses spanning 2 to 3 years. That's a long time to wait if you want to start analyzing data for your own business or investments. You need an alternative—a program that is much quicker. The book's program offers an alternative path relying more on modern software than math skills. Which is better, this program that uses software and bypasses math or the route of taking several math courses before you start analyzing data? A fair question, but since the answer differs from situation to situation there is no one answer. Here is why. Often your specific business area knowledge plus the math you have already taken plus the recommendations in this book will be enough. Nevertheless, more math can't hurt. The more you analyze data, the more you will realize that. Advanced math helps you do better analysis—no one is arguing the point. However, data analysis generalists require advanced math. Specialists in a business area who want to model with data can often bypass a lot of math. And that's the bottom line on math.

You need some math, of course, to understand models and run the software. Overall, the program requires algebra II. However, if you have some calculus and statistics you will understand the program better. Here is a diagram to illustrate. The diagram shows two possible paths that you could take. The right-hand side bypasses several math courses and uses a different set of modeling programs. Both paths lead to a model. Both use software at the end. What is the difference in the final model? There may be no difference at all. The techniques used in this book are accepted and used by mathematicians and statisticians everywhere.

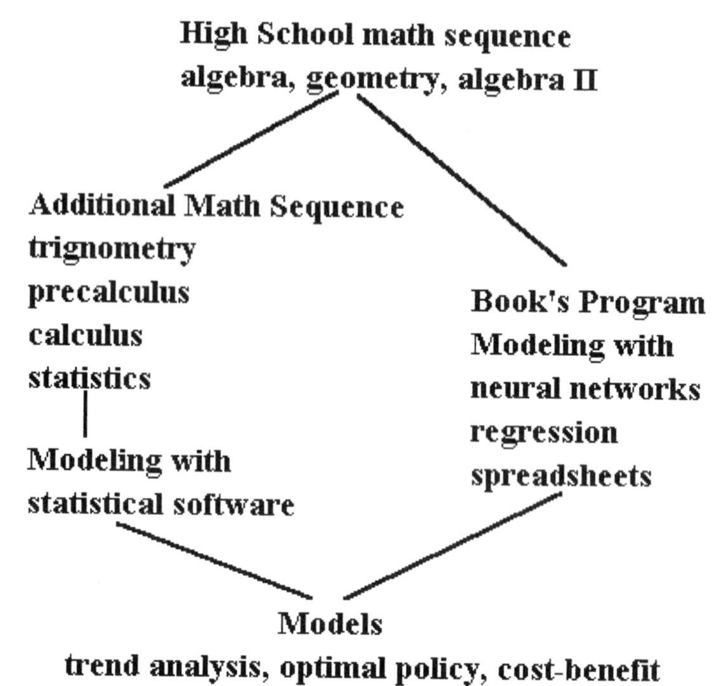

These techniques are neither magic nor "dumbed down statistics." However, formally trained, statisticians will often feel more comfortable using statistical techniques. These techniques come with accompanying analysis like significance tests, analysis of residual errors, partial and serial correlations and collinearity concerns. Most of which is pre-modeling phase data reduction or post-modeling validation. Statisticians learn this approach in college statistics courses and feel comfortable with it. The actual construction of a model from data does not require this pre-modeling data reduction or post-modeling validation with mathematics. A legitimate method of choosing input data is to be an expert in the area and to understand correlation and cause-effect relationships between data sets. A legitimate method of model validation is for an expert to look at its outputs and verify that it indeed simulates the actual process that it purports to model. Experts in a business area can bypass a lot of math when analyzing their *own* data if they don't have an advanced math background. That's the key point.

How Difficult Is It To Model?

How tough is modeling? There isn't any short answer, yet we can put the question in perspective. The next table places several modeling techniques on a 1-10 scale with several math courses and computer skills. Now you have a personal perspective on the difficulty factor. On this scale of difficulty of standard high school/college math classes or standard PC software (1 = calculator, 2 = Windows 95, 3 = high school algebra, 7 = calculus, 10 = college engineering calculus) the method uses five techniques that all are ranked between 1 and 3.

Level of Difficulty of Data Analysis Methods

Math Class or Software	Difficulty	
calculator	1	
correlation, regression	1	(part of method)
Windows 95	2	
Databases	2-3	(part of method)
Algebra I	3	
Spreadsheets	3	(part of method)
Neural Networks	3	(part of method)
Algebra II, Trigonometry	4	
Intermediate C or C++	5-6	
Statistics	6	
Calculus (non-engineering)	7-8	
Calculus (engineering level)	10	

You can apply math modeling to your data—believe me. If you are still skeptical—read on!

Even 10-Year Olds Can Make Math Models

To test this program out I had my 10-year-old son Cameron learn the correlation, regression and neural network techniques. Then he ran them with the same data used in later chapters. As the following table shows, it did not take him long to learn.

Time for 10-year old Cameron to Learn and Run Modeling Software	
Regression	1/2 hour
Correlation	1/2 hour
Neural Networks	3 hours

Of course he can't do any actual business analysis from start to finish. He doesn't have any business experience. What he did was learn to run the modeling phase. He learned what keystrokes to enter on a calculator. He learned how to prepare data files for input and what Windows commands to use. Finally, he learned what variables to set in a neural network program. Still, the point is that he could do this and in a short amount of time. You might ask: "So what, could his model actually analyze data—did his answers make any sense?" The answer is yes! He was given the data from a bakery example in a later chapter (a file of data with 360 rows and 5 columns). After he ran the neural network and got a model it was compared to other models based on profitability. The next table shows the results. His results are quite good. Still, his results reflect the robust nature of a neural network more than anything else. Although the name neural networks has this foreboding, impossibly difficult, modern-AI-technique sound to it—they are not hard to run at all. Sometimes getting good results with them is almost too easy to believe. There are many cases with users that had no more math background than high school algebra. Still, they were developing good models with neural networks soon after taking the neural network software out of the box. But more about that in the neural network chapter.

Bakery Modeling Scenario

	Model	Profits ($)	
	Normal business judgement	29,000	
	Multiple linear regression (statistical method)	73,000	
**	10-year old Cameron's Neural Network Model	73,000	**
	Neural Network (Ph.D. running it)	84,000	

Cameron essentially ran the program with vendor-supplied default values for the parameters and still developed a good model. Compare his results to the multiple linear regression model (a mainstay of statistics)—the same profit figure. However, Cameron's model was better than the multiple linear regression model because it led to fewer days when the sales were underestimated. This would mean that a bakery owner would choose Cameron's model over the multiple linear regression model because it would mean fewer unhappy customers. His results are not much worse than mine and I tried a lot of different options and ran the computer for many hours more than he did. (One disclaimer: his results were outstanding for this type of problem because a neural network is ideal for this type of prediction problem. A 10-year-old wouldn't do as well for some other examples in the book, because they required more integration of modeling and business savvy.) Still, the key point remains, the programs that do the actual modeling are not difficult.

Just for kicks, I asked Cameron to rate the difficulty of the modeling methods against things in his world. (He hasn't taken any of those high school math courses since he is in the 5th grade. However, he has learned some algebra and geometry at home with Mom and Dad). He ranked them against three video games for the Sony PlayStation™. Easiest was 1, hardest was 5. Easiest at 1 was "Twisted Metal™." (Cars drive around smashing into each other and shooting each other to pieces. Last car running—wins.) At 2 was calculating correlations and regression on a calculator. At 3 was "NFL GameDay 98™." (A good video football game with lots of different formations and plays.) At 4 was running neural network software. Hardest at 5 was "Final Fantasy VII™" a very popular role-playing game, with lots of tables of weapons, attack points, magic spells and a large number of playing options. Draw your own conclusion.

Guide To The Rest Of The Book

The next chapter is basic to all that follows; it discusses the progression from data to model. Models don't have to be mathematical formulas. Other forms are computer programs and sets of rules. You have options. The third chapter goes into more detail on some continuing examples and shows how profitable better analysis can be. Then comes a series of chapters of two types: 1) specific modeling methods or 2) general topic discussions. You can read them in any order after chapter 3 although in a couple of places you might have to refer back to the statistics chapter. There are real world case studies and examples throughout this book. It is here that most of the learning comes from. These examples will give you ideas about how to view your own data. They will give you ideas about which software techniques to use. They will give you ideas about what you can expect to get from data analysis. Your own data and needs are unique; you will have to come up with ideas on how to analyze them. It's the ideas in the examples that count most. These ideas will get you thinking about your unique problems. Author John Steinbeck put it best many years ago: "Ideas are like rabbits. You get a couple and learn how to handle them, and pretty soon you have a dozen."

Chapter 2

From Data to Model
From Model to Plan
A Program for You

Have you ever faced a mass of data: a table, a stack of sales slips, a ledger full of figures or a spreadsheet and knew that it was useful but had no idea what you could get from the data or where to start trying? Data is everywhere; often we must analyze and understand it. Large corporations employ small armies of computer specialists, statisticians and MBAs to analyze their data. Small business persons and individual investors don't have these resources; they have to analyze their data themselves. Data is king and everyone is more valuable if they understand data analysis.

Math classes are not necessarily the answer. Oh, the more math you have the better you will analyze and use your data. Nevertheless, people with only high school algebra and some computer skills can do many practical and profitable data applications. Modern software tries to be user friendly and that means "as little math as possible." Many modern software packages now handle much of the math analysis for you. Even more, some data analysis problems just don't require that much math if you formulate your problem the right way.

Should you sign up for more math classes at your local community college? Should you rush out and spend hundreds of dollars for new software? For right now, neither. This chapter shows you where to start, how to define your goal and what direction to start in. Several examples are given. These are real-world, real-people, won't-make-the-nightly-news examples. These are not textbook examples of planets orbiting the sun, analysis of trade with Mexico, economic analysis of the whole U.S. economy, or predicting the environment in

2020. These problems are for everyday people: small mail-order businesses, middle school teachers, bakery owners, and small franchises.

There usually isn't any math course in high schools dedicated solely to data, trend analysis, data handling, data fitting or prediction. Yet this area is more important to most American adult workers than geometry, trigonometry, or calculus. Managers made all business decisions fifty years ago. They were expected to watch their area of responsibility and decide for their company. Occasionally dictatorial, often middle-aged to elderly and overwhelming white male, these managers were one-man decision machines. World War II started a change to making decisions by more quantitative methods and this required management staffs. The computer age has accelerated and improved the process. Today, many companies make decisions by using small teams composed of several levels within the company: line workers, sales agents, engineers, managers, planners and facilitators. A keen detailed understanding of data analysis can lead to much higher paying jobs: MBAs are the quantitative experts in business and they have been in high demand for years. However, you don't need an MBA to make quantitative decisions in your present job. This chapter will go into the basics and then later chapters on statistics and neural networks will give you practical tools to use.

How You Use Models

Often we need to consider how to predict a quantity. This quantity is an unknown to us. Still, we know the major factor(s) that determine it and we need to use this information to predict the quantity. We know the numerical value of the cause. We want to know the numerical value of the effect. To do this we use a function that relates cause-to-effect or data to unknown quantity or independent variable to dependant variables. Mathematical notation usually uses letters like "f" or "F" for functions and "x" and "y" for variables. Functions end up looking very abstract like: $y = f(x)$. But don't be put off by this—it's simply shorthand notation. We can use more descriptive notation if we want. For example: a function that relates a student's 5th grade math grade to her 6th grade math grade. Instead of the mathematical forms: f, F, x, y, etc. we might use "grade5" to be a number between 0 and 100 that represented the average math grade and the name "model" to represent the mathematical function that relates "grade5" to "grad6_pred" the predicted academic math class grade she could expect (the actual grade could vary from this of course). In mathematical form this could be: *model(grade5) = grade6_pred*. Other examples might be like a graduating senior student estimating how much salary she might earn with her soon-to-be B.A.. Not all students are offered the same salary; companies differ and most companies take into account other factors besides the degree like your position in the graduating class. If our student knew a formula (which is the same as a function) she could estimate her starting salary by substituting in her rank in her graduating class. For example: *predicted_salary = salary(class_rank).*

Previous examples were for the most part single-variable functions—one input, one output or one independent variable, one dependant variable. Functions or formulas with more than one input (independent) variable are common. For example a baker could estimate sales of donuts for the next day by a formula like: *donuts = predict_donuts(day-of-the-week, temperature, rain, shopping_season).*

Prediction also allows us to choose optimal plans. The people that want to start a franchise in a local area might have several possible locations. Each location might have a different expected income, a different population density and a different rent. They need to estimate the expected income from each site and then subtract off the different rents to find their expected profit. Then they can determine the maximum potential profit by comparison. This can become a complicated multi variable procedure in practice. The form of the prediction device has many names: it can be called a model, a simulation, an evaluator, a trend predictor or simply a function.

How You Get A Model

So far, we have seen several practical examples where we can find a quantity that will be very profitable or satisfying for us to know. These were like how many donuts to bake tomorrow to avoid waste and maximize sales, or the expected grade in 6th grade math. Profit and satisfaction came from just substituting known numbers into a formula and cranking out the answer. Is data analysis that simple? Not by a long shot. The problem is *not* substituting in a function and cranking out the answer. The problem is *finding that function* in the first place. The function is the unknown in data analysis and trend analysis.

Finding a model for our data requires several steps and new concepts. First, we need to look at data, generic numerical data. What is data, the role of uncertainty and what needs to be done to it to make a model. Data is basic to us. Often it is all we know (or will ever know) about our situation. We use data to develop models, then we use models to predict or optimize or plan with. Second, we must look at functions: there are an infinite number of functions out there, how do we know which one should be our model? What criteria do we use to choose one function out of that multitude of possibilities? How do we find the correct one once we have chosen the criteria?

Developing Rules From Data—Mail Order Case Study

Ronald has just paid a fee to a mail order company that allows him to sell their merchandise in his area. He knows that cold canvassing via the mail is his only option so he wants to get a list of reliable customers for future mailings.

Mailing out flyers to the whole area is prohibitively expensive. He needs to be smarter than that. The company did a trial mailing of 10,000 in his area as part of the franchise agreement. So he has the results of that. Mass mailings seldom return more than 4% and he probably has to get a 2% or 3% or higher return to pay mailing expenses and make enough profit to live on.

Ronald has recently retired and started a mail-order business.
He needs at least 2% positive replies to make money.
How can data analysis help him?

What should he do? Ronald has some ideas but doesn't have a workable plan. Listen to his situation. "I move a lot slower than I used to. Recently I retired but my pension isn't that much and I would like to stay active. Mail-order offers me the chance to work at home and still have enough income to be independent. Still it is risky—if anybody can make big profits like those ads lead you to believe—then a lot more people would be doing mail-order. I have gone to conferences and conventions and I didn't meet many rich people."

"There was an article in Time magazine about the science behind mail-order. Lots of obscure mathematical stuff like regression analysis and market forecasting. At the conferences outside speakers put up graphs and tables and diagrams of how to choose a good mailing list for a new product. It's all mathematical and computer analysis. Now I can handle a PC and I am willing to work long hours and I can learn new software. But if my lack is advanced math then I am in trouble. It's too hard physically to get out to the local college for night classes and my high school math is very rusty by now. Is there software that can do some of this market forecasting for me? Can I get software that will help me choose a profitable mailing list instead of high powered mathematics?"

Background: his mailing is the company mail-order catalogue for age-specific children's gifts. The company sells him the catalogue. He mails them to the customers then when his customers order something he gets a percentage. The mailings are not much work, but he does bear the risk of losing money due to insufficient sales to cover his mailing expenses. He can easily get mailing addresses from companies or on a CD-ROM.

Analysis: What types of people would be his best customers? At first glance: people with kids; because kids want presents, have friends that invite them to five or six birthday parties a year, which requires that many more presents. But all he has are mailing lists of names and addresses provided to him by the parent company. The following table is a sample.

Raw Data

Mr. Sam Smith
604 Elm St.
Hightown, VA 73210
(yes-bought)

Mr. and Mrs. Clyde White
11073 N. Bloom Apt #102
Hightown, VA 73216
(no)

Ms. Jane Doe
712 Sparrow Apt #6A
Hightown, VA 73214
(no)

Mr. and Mrs. Joe Green
107 Eagle Rd.
Hightown, VA 73217
(yes-bought)

He can tell something about the people on the mailing list by looking at four parts of the address: title, sex, apartment number (if it exists) and zip code. By the marital title and/or gender you have a good idea if the addressee is a married couple or a single (could be separated or divorced) man or woman. By the apartment number or lack of it you can tell if the addressee lives in an apartment or in a more expensive detached home or townhouse. By the entire address you can get an estimate of the average housing cost or apartment rent for that area. That is a rough estimate of household income.

Ronald is not sure who will be the best customers. Women buy more kids' toys, but they also shop in retail stores more (and better) than men. Bachelors are probably duds. They probably don't buy many kids' presents at all. Yet divorced or widowed men with kids could be a profitable source. They may not have much time to shop or interest in shopping. They might like the quick easy method of mail-order. But who knows until you analyze the data. In addition, home addresses are assigned rough estimates of worth or monthly rent as a rough indicator of household income. This gives several more categories. By using a database program that totals up all people on this list by different categories he can form a table with percentage positive (they bought something) responses.

Processed Data

Mail Order Catalogue Purchase Rate (%)

	Estimated Value of Home		
	$100,000	$150,000	$200,000
Married/House	2	6	2
Single Woman/House	1	2	4
Single Man/House	2	9	12

	Estimated Rent(per month)		
	$500	$700	$900
Married/Apartment	1	2	3
Single Woman/Apartment	1	2	2
Single Man/Apartment	<1	1	1

From this table he can now generate three rules that allow him to choose categories that return 4% or higher positive returns. These rules are his model.

Model

Mail to married people in houses in areas where the average home costs over $150,000 but less than $200,000.

Mail to single men in houses in areas where the average house costs over $150,000.

Mail to single women in houses in areas where the average house costs over $200,000.

All it took to do this analysis were a few logical searches with a standard database program on a PC. The rules generated from the table will save Ronald thousands of dollars in potentially wasted mailings. We will discuss database programs in a later chapter. They can convert raw data like the mailing lists into the tables of results and the final logical mailing rules that we saw. Rules can be a model too. If you wanted, you could convert these rules to math functions. Still, people are comfortable with rules. They are easy to use. So often a model is a set of rules.

Predicting with Data—Bakery Inventory Case Study

Maria owns and operates a little bakery and dining area in a local shopping mall. She can only bake during the early morning hours and needs to have an ample supply of donuts for the rest of the day. She can bake so many extra that she won't run out but then she throws away a lot and the wasted ingredients cut her profits down. She wants to be able to predict in advance how many donuts she will sell.

Like Ronald, Maria has some knowledge about what trend prediction and data analysis can do for her. Also, just like Ronald she hasn't any idea where to start. She says: "The bakery chain I used to work for had a computer link to corporate headquarters. Every night the manager would input some predicted weather data from the local TV station and send it off to headquarters. Later he would get suggestions back from headquarters about how much business to expect."

"I knew that they had some MBA types doing this market forecasting stuff. Met them once—just kids really, late twenties. Probably smart enough, but neither of them had any experience behind a counter or oven. And here they were telling us how to do our business. But they were awfully close most of the time. We almost always had enough donuts to last the day. Never seemed to waste too many either. That is what I need to be able to do; but I can't afford to hire an MBA consultant. I am on my own. What can I do? I don't have an MBA or even a college degree; high school math was easy enough but I don't have calculus or statistics and don't have time to learn them now that I am running my own business. Is there any software that can do that MBA-type market forecasting for me?"

The main factors are the day of the week (more customers on the weekends and most diets start on Monday and dieters' resolve starts fading later in the week). The shopping season (Christmas season, holiday weekends, summer and winter gets different amounts of shoppers). The weather (temperature and rain, colder weather brings out more pastry buyers, rain discourages people from coming to the mall). Also, there is an unavoidable randomness due to factors like a few local companies that once a week or so hold large conferences and order 200 or more donuts for the attendees. Maria has the sales record and weather reports from the previous year so she has hundreds of entries like the following table.

Maria just started her own donut shop in a local mall.
She only sells fresh donuts—but donut sales vary widely from day to day.
Can trend analysis predict the next day's sales?

Raw Data

Day of Week	Temperature	Rain	Season	Donut Sales
M	40	0	winter	1200
T	28	1	winter	600
W	36	1	winter	1200
Th	40	0	winter	2100
F	46	0	winter	2200
S	20	0	winter	2400
Su	25	0	winter	2000

She knows that this is important. There are useful patterns in the data that she could profit by knowing. She would like some rules to follow like Ronald's mail order rules or even better a function or computer program which would do it for her. She knows that there is probably some complicated math function that relates causes: *day-of-week, temperature, rain, season* to the effect: *donut sales* but she doesn't know it. However, there is a way she can use her data to build a model that then predicts for her—a neural network computer program.

A neural network is like an explicit function (e.g., $f(x) = 2x + 4$) in as much as she can predict with it. However, the exact form of the function is hidden within the neural network and is quite cumbersome to state. Still, Maria can predict sales with this neural network. She will be able to input the known day of the week and the predicted weather into the program. Also, she can input a numerical rating of the mall attendance during the season. Then she can run the program and it will output predicted donut sales. In this case shown in the next diagram she wants to predicts tomorrow's (Wednesday) donut sales when the weather report calls for 38° with 1" of rain and the season is early November. The neural network estimates that sales will be about 1300 donuts. She can then adjust her morning baking and save money. It will look like: day-of-week, temperature, rainfall, season going into a neural network program which outputs predicted donut sales.

This might seem like a little bit of magic. Indeed, they often refer to neural networks as "black boxes." This is since you don't know what is going on inside them, but your answer comes out the other side. Neural networks are part of the new generation of Artificial Intelligence (AI) software. Later, we devote an entire chapter to these neural networks; they work like the human brain, learning by analyzing data and creating patterns. Neural network expert and author Jennette Lawrence has this to say about this new type of software: "Neural networks, on the other hand, are actually able to learn by creating their own internal representations of reality based on raw information given to them."

Model

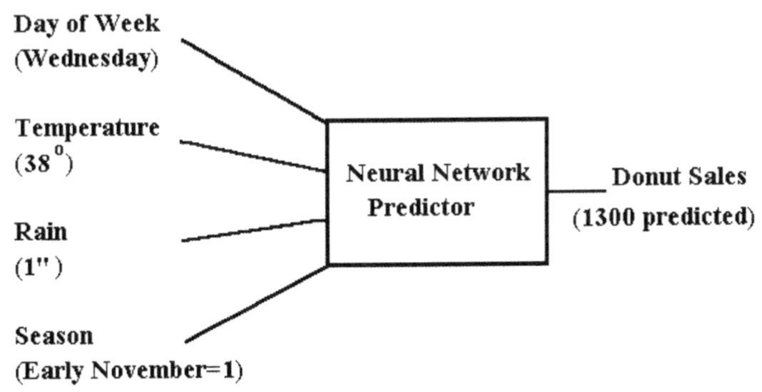

And neural networks work in the real world: They have designed neural networks that are 90% accurate in predicting the actual sale price of real estate in several tests. More later about neural networks. However, for now you just need to realize that such things exist and you could use them.

Models That Fit Best

Data analysis is not about collecting or archiving data; we don't want to be data historians. Data analysis is about using data from the past to predict what the future will be like. Data analysis is prediction, planning, optimizing and solving. How will this 5th grader do in 6th grade, how much will the S&P 500 change if the new interest rate increase goes into effect? How much will it cost our company to carry out this new governmental regulation? How much will the postage cost if we mail enough catalogues to attract a thousand new customers? How long will it take us to sell this inventory? Can I buy this new franchise, set it up in my home and make enough profit to live on? Will it be profitable for us to add a garage to our house if we want to sell it?

The government uses this type of analysis all the time. If we want to be informed citizens, we need to understand how they are doing this. The government is in the business of spending money (our taxes) to improve the lives of its citizens (us). There are a million ways to spend money. No matter how many billions or trillions of tax dollars there is never enough money to do everything. Tradeoffs are necessary: They fund a program, they cancel another program. How much good will be done by funding a program; how cost effective is the proposed program, how much "bang for the buck?" Will a million dollars extra spent on education lower crime more than a million spent on extra street cops? Will paying extra teachers to lower the average class from 24 students to 20 students be worth it? Or could the money be better spent on extra student counselors or improving school infrastructure? Like it or not, quantitative methods are required. We can't plan wisely by letting some single manager make summary judgements based on experience. We need supporting data, analysis, revisiting original plans, and mathematics. Data is everywhere and data analysis is required. Data analysis, trend prediction, regression analysis, modeling, cost-effectiveness, tradeoffs and their own peculiar vocabulary and concepts have become the language of modern business and government.

One overarching fact: we seldom know all the causes of an effect. Our knowledge of a situation is usually limited, sometimes severely. Many factors affect a 6th grade student's performance in mathematics. Some of these factors are impossible to measure. For certain, a good predictor of 6th grade success is the 5th grade. Still, scores on standardized math tests are almost as good a predictor (especially at the more advanced academic and honors level). Less so are factors like the quality of the textbook, the methodology and skill of the classroom teacher, the social circle that the individual student hangs out with, the amount of support and help that the student gets at home, intangibles like the student's interest in math and others.

Two students both with an 80% average math score in 5th grade might differ considerably on other factors. One might have an improving home situation that leads to a higher than expected grade. The other's home situation could be worsening and leading to a decline in grades. Also, because of interest and peer pressure one of the two students might have a much higher mathematics aptitude but not have studied as much in 5th grade. Maybe because math was "not cool" or considered "hard," which would lessen her relative advantage. This is all obvious. But how can we ever expect to predict a student's academic performance with all these unmeasurable unknowns? We can but it isn't easy; unknown factors disrupt the prediction process.

It all affects our predictive ability like the following. Suppose we have a model (e.g., a function that takes 5th grade performance, average grade (0-100), and when we substitute that in, the answer is expected 6th grade performance (0-100)). "Expected" is the key word. If we have four students with identical average grades in 5th grade the model would predict exactly one grade for each of them (say 80%). Now it's likely that none of the four would achieve exactly 80% for the 6th grade year. Instead, we could expect that one might do poorly, say 70%, a couple might hover around the expected value, say 78% and 82% and the last might do better, say 85%. The causes might be that the 70% student simply had a bad year—his parents were divorcing and he couldn't concentrate on his studies that entire year. The 85% student might have been an underachiever during the 5th grade year and finally matured enough to realize her potential in the 6th grade. It is very possible that she could have been recognized earlier by her outstanding performance on mathematics aptitude tests if they designed a model that included both 5th grade scores and math aptitude test scores.

Therefore, our data will look like this: students with exactly an 80% average in 5th grade will have a range of grades in 6th grade math. That holds for all other grades also. A model will predict a 6th grade math score. Still, there will be some uncertainty based on the other unmodeled factors. No surprises here. We must live with some uncertainty: grades, Dow Jones averages, Super Bowl winners, presidential elections and life itself are uncertain. Still we want to do the very best we can do. How is this possible? To start with we need to define "best."

We use the word "best" in many senses: Michael Jordan is the best guard in the NBA. "Best" here is in many senses: scoring, defense, leadership, and certainly salary. Scoring and salary are numbers and can be compared to other NBA guards quantitatively. Leadership and defense are more subjective and therefore more debatable. "I want the best possible price for my next house." Here the prospective home buyer is probably talking about best in the sense of the least cost. "That dress looks best on you." Here is a friend's judgement that of all the clothes that she has just watched you try on this particular dress looks best. Here best is a composite judgement. Overall, when you weigh together all the ways that the dress is evaluated (style, color, pattern, sleeves, etc.) this dress is best overall.

A "best" model will combine features from each of these examples. First, we use the terminology "fit" in data analysis; it means an overall weighting of the match of the model to the data. It is like the composite judgement for the "fit" of a dress. A fit that is close but not perfect everywhere is usually preferred to a fit that is perfect in several places but badly off in others. Second, like Michael Jordan's quantitative categories of scoring and salary we insist on a quantitative measure of goodness of fit in data analysis. Third, like the home buyer who wants the minimum cost we insist on a model that gives the minimum variance from the actual data.

"Best" fit is a worthy goal but we are talking about math functions here. How will this work? Since this idea is basic to regression, statistics and neural networks, we need to examine this before going on. The easiest way is with a regression example.

Regression analysis allows us to determine mathematical function models. This in turn allows us to predict, plan, optimize and solve our everyday business problems. Regression is easiest to explain by showing the concepts of "better" fitting function and "best" fitting functions. Suppose that we have data taken at three values of the independent variable x. Say x = 0, 1, 2 and the corresponding data or dependent variable values are 2, 4, and 6. Let's look at how two different functions f(x) = x + 4 and g(x) = 3x + 1 "fit" this data. Our measure of "fit" is a residuals function. A residuals function is the difference between the values predicted by the fitting function and the actual data. Each difference is squared and then all the squared differences are summed together to give the residuals function.

Therefore, g(x) is a "better" fit than f(x) to the data. It turns out that the "best" fit to the data is the function 2x + 2. They call this the "linear regression"

Which Function Fits Better?

x	0	1	2		x	0	1	2
data	2	4	6		data	2	4	6
f(x) = x + 4	4	5	6		g(x) = 3x + 1	1	4	7
f(x) - data	2	1	0		g(x) - data	-1	0	1

residuals $= 2^2 + 1^2 + 0^2 = 5$ residuals $= (-1)^2 + 0^2 + 1^2 = 2$

"worse" 5 > 2 "better"

function. We discuss it in more detail in the statistics chapter. A linear regression solution is quite easy to obtain from spreadsheet programs or even good hand calculators. It is the most widely used fitting function of all. The main problem with linear regression is not calculating it. The main problem is whether the data warrants using a simple math function, like a straight line, as a math model for the underlying process.

Prediction With Regression—Middle School Counseling Case Study

Every year millions of 5th graders move onto 6th grade and middle school. These students were all taught in one room by one teacher in the 5th grade. In 6th grade classes become individual classes in different rooms taught by different teachers who specialize in different subjects. Classes which they taught to everyone are now broken into instructional levels like: basic, general, academic and honors. Cynthia, a middle school counselor, has the task of placing the incoming 5th graders in one of these four levels of math classes. Her recommendations are not binding, parents can ask that their children take a different level. Usually, they accept her recommendations for the basic, general and academic levels. (Honors classes usually have stricter predetermined requirements, like top 10% in math aptitude testing and A or B averages, so Cynthia doesn't place students in honors math classes). At first glance, Cynthia's job would not seem mathematical. She counsels kids. That seems to require teaching ability, understanding of psychology, teaching experience, a good understanding of the regulations and a good deal of experience handling difficult children. But mathematics, why mathematics? Because data is everywhere. And you can't do many seemingly non-technical jobs or advance in them if you can't handle data.

One trend that Cynthia hates is that sometimes parents of a bright, intelligent 6th grade girl override Cynthia's recommendation. They enroll their daughter in general math instead of the more challenging academic math. This has happened hundreds of times over the years and her best efforts at convincing dubious parents usually fails in this case. These parents are usually quite intelligent themselves, but they fear that a harder math level would be more stress for their daughters. Many of them have asked Cynthia to be more specific about their daughters' chances in academic level math. During parent-teacher conferences Cynthia used to rely on just personal judgements and observations. Parents would like to see a predictive method used. A predictive method could be based on their daughter's grades, standardized test scores and comparative grades from former students with identical qualifications. That is a reasonable desire.

Listen to Cynthia tell her problem. "I took this counseling job to help kids. Assigning kids to different math class levels is my hardest problem. The 'avoid math' trend starts in middle school; most elementary school girls like math and do well.

Cynthia is a middle school counselor.
Every year many intelligent 6th grade girls do *not* take challenging math classes.
What can data analysis possibly do to help this?

But when they get to middle school, they start dropping out. Yet, there isn't that much social pressure to avoid math like there is in high school. Math is still 'cool' in middle school—well, at least cool enough. Overprotective parents are my problem. They listen to me, they smile, they say that they want to think it over. Then later I find that they enrolled their daughters in general math instead of academic level math. They are nice enough, but they are afraid of stressing their daughters out. Still, the girls that I recommend for academic level math can pass. I know it, there is a lot of data to prove it. Lots of girls with the same 5th grade scores have passed academic level math in the past. I just need to convince their parents. A good prediction method that parents understood would help a lot. Many parents have asked if we have a prediction method like the one colleges use with high school grade point average weighted in with SAT scores. We don't. Maybe we should."

So Cynthia set off to do just that. She had many years of historical data like the following table.

Raw Data

	5th grade math average	Standardized Math test scores	6th grade Academic math average
Samantha	89	43	86
Lynn	95	51	93
Ryan	87	47	89
Carol	80	40	75
Matthew	74	50	80

Now there are always differences between individual students. Some simply like math more and spend more time studying it. Some are hyperactive and physically can't sit still long enough to study. Some are having troubles at home. Some are going through a parental divorce. These are random factors affecting performance that don't show up in the math average or the test score columns. Cynthia can never predict perfectly. Still, she can predict better than now. Besides, she will have evidence to support her recommendations. This will help those dubious parents. What she needs is a rule or function that weighs together 5th grade math grades with standardized test scores into a predicted grade for 6th grade academic level math. This can be done; colleges do it all the time. They predict applicants' chance of success; often combining an applicant's high school average and their SAT scores.

To do this she takes all the available data like the above table then she runs it through a multiple linear regression statistical routine and comes up with the following equation.

Model

$$grade\_6 = grade\_5 - 3 + .5(test\_score - 47)$$

(This formula predicts like this. Sandy average 86% in 5th grade math and tested out at 49 on the standardized test. The formula predicts that her 6th grade academic math score will be: grade_6 = 86 - 3 + .5(49 - 47) = 84. Even allowing for some random factors in the prediction Sandy can easily handle the academic challenge and Cynthia is justified in telling her parents so.)

Cynthia sees a lot of truth in this equation. First, the average test grade of the academic level students is 47. For half the students (47 and above) it will predict slightly higher grades than a student with the exact same 5th grade average but a test score that is lower than 47. Second, the average grade of all students has been about 3% less than their grades in 5th grade. Simply put: academic math is more challenging and 6th grade is tougher than 5th grade—nothing surprising here. Also, the past grades weigh more heavily than the test scores; which has been historically true but has shifted some in the past few years due to "grade inflation."

Now, Cynthia has a good predictive tool. It is more exacting than her previous advice which was based on looking over the data tables. Moreover, she can show it to dubious parents. It will reinforce her opinion and it is demonstrably accurate. Cynthia can expect many more bright, intelligent 6th grade girls to be taking the more challenging academic level math in the future. Cynthia won't profit financially from this like Ronald or Maria. Still, it will enhance her counseling; it will turn out more girls who don't shy away from challenging math. A job well done.

Conclusions

We have seen how you can use your raw data to develop a model that helps you make decisions. There are different forms, different goals and different methods. Always, you need data and a goal like: best people to mail catalogues to, a way to predict math grades or how to predict tomorrow's donut sales. We also saw the outcomes that you can expect: a handy formula, an easy to use set of rules and a computer program. The latter chapters in this book show you how to do all of this analysis. They expand the discussion of these examples and take you through step-by-step methods to make it easy for you to apply these techniques to your problems. And that is your goal after all.

Chapter 3

Levels of Data Analysis

Alternative Methods For Non-Mathematicians

Have you ever faced a problem involving lots of data and wondered if you could make something useful of it? Have you wondered if you have enough math? If not how much would you need? Do you have any options? Are there different types of analysis, some simple, some not?

The previous chapter explained what kinds of results you could expect from data analysis. This chapter explores alternatives and these different analysis options lead to different levels of analysis. We aren't all Ph.D.s; we all can't use the most complex, the most math intensive techniques. Nor do we have to. Often simpler mathematical techniques give as good an answer as the most complex statistical calculations. Often you might be satisfied with a less-than-best model. There are tradeoffs and alternatives. You will see several in this chapter.

This chapter is also about your math and computer background: the alternative methods of analysis are keyed to different levels of math and computer backgrounds. You will see that much analysis can be done with little math. Also you will see where the math requirement gets deep quickly. Data analysis at many levels can be done with high school math. Still sometimes you might step off into deep water—advanced nonlinear regression requires calculus, statistics and numerical analysis—that is the deep water.

We are going to revisit examples from the previous chapter. There we saw what data might look like. Also, we saw what prediction tools we might derive from that raw data (e.g., rules for mail order, an equation for predicting grades, and a program that could predict tomorrow's donuts sales). This chapter will find us looking at many different levels of models, rules and equations. Each will be better than its predecessor and not quite as good as its successor. One big question will be: how does better analysis translate into success, like more money? Each example tries to answer that question in some way.

What do we mean by levels? The same problem can be modeled with a few general methods. Consider the donut bakery example: one predictive model is that the average daily sales are 1500 donuts; this isn't much help or very profitable. Another level might be that there is an up trend from Monday to Saturday where Monday sales average 1100 and Saturday sales average 2000. Now, this is better, some adjustment to baking can be done based on this predictive model, but we can do far better. Perhaps like Monday sales average 1100 and each successive day's sales add 150 donuts to the previous day's sales. Also, winter sales average 10% higher than spring and fall sales and 20% higher than summer sales. Now, you can do a lot with a two-sentence rule like that. Still, you can do better. The highest level might be a computer program. This program might take the day of the week, the season, the predicted temperature and rain for the next day and give an accurate prediction of how many donuts your customer base is likely to buy. Four levels: one simple rule based on a simple average, two higher level rules that quantify trends that are based on spreadsheet analysis and one complex program that takes into account all trends and bases its answer on complex mathematical calculations. Four levels of profit too: baking the average sales every day will leave a lot of surplus and shortfalls, you will both waste a lot of money on slow sales days and lose potential sales when you don't bake enough to last through the high volume sales days, not a good method. The next two rules will profit more, there will be less shortfall and less waste if you use either of them. The largest profit and the least shortfall will come from using predictions based upon the most complex computer program method.

The most exacting might be an equation like: $donut\_sales = 1000 + 145*day\_of\_week + 25(60^o - temperature)$. This wouldn't be any better at prediction than a computer program but it would give you more insight into what factors were affecting sales. For selling donuts an accurate prediction is enough. Yet for predicting math grades when you are counseling students having an equation plus the additional insight would be better. Different methods are best for different problems. Although, it might not have been obvious, the mail-order problem was the simplest of the three. The bakery and math grade examples were about the same order of difficulty except that for this "insight" issue. You need this insight for many applications in the social sciences. Social science applications usually involve peoples' welfare. Statistics are widely used in these areas because the solutions can be checked for statistical accuracy and any inaccuracy can be pinpointed better. Neural networks can often be used for straight prediction. Cynthia needs an explicit function; Maria just wants to estimate tomorrow's sales. Cynthia needs statistical analysis; Maria can get by with the neural network approach. Statistics require more math background than neural networks. So if you don't have a calculus or statistics background, then you should consider neural networks.

Mail Order Case Study

Ronald has taken all that raw mail order data that the franchise provided as a sample. Next, he put it all into a commercial database program. (He didn't even have to buy a database program. They often provide good ones as part of the resident software when you purchase a new PC. I have been using database programs on my home PCs for years. I never bought one. I just use the ones that come with the computers.) As we saw before he can group together addresses by three different parameters, all of which tell something about the addressees as a potential customer. The three were: title (information about marital or living status), apartment number (tells if they live in an apartment or a more expensive townhouse or single family home) and location (both zip code and street addresses). The database system allows him to sort the data based on these fields and derive composite information like the following. What percentage of married couples ordered merchandise? What percentage of people that live in houses ordered? Recall that he probably needs a positive response of 2% just to cover mailing costs which he has to pay for entirely. When he totals up the positive replies from the mailing test list that the company sent out the positive rate is only a bare 2%. Ronald needs to look for identifiable subgroups of people that will buy more than the minimum 2% of the time.

Maybe you think that this is a waste of time and that Ronald should just shotgun out mailings. If so then consider some following comments from US News and Reports (Dec. 8, 1997) concerning the direct mail business: "... direct mail is a booming industry because it targets us with scientific precision" and "Driving the explosive growth in the direct-mail business is a vast data-collection and data-crunching network." Just consider those phrases: "scientific precision" and "data-crunching" they cry out that data analysis is taking place on a large scale in this business. Ronald may have only a small piece of the direct mail business. Still, he had better be up to the task—it is a hard sell—US News and Reports commented that 95% of the people throw away their catalogues when they receive them.

Ronald's problem is getting the largest rate of return possible to maximize income. The company says that the average order size is $80 of which his share is 25% or $20 and the mailing cost is $40 per hundred. So 2% orders is the break-even point. Here is a detailed cost analysis of the problem: though he can only handle 5000 orders of an average profit of $20 apiece, he must bear the cost of mailing which is $.40 per addressee or $40,000 per 100,000 mailings. If he mails out 200,000 letters to get those 5000 orders, it costs $80,000 leaving a yearly income of only $20,000; however, if he can get the same number of customers in only a 100,000 mailing the income is now $60,000. Crucial to his yearly income is that rate of positive return.

Since the entire list doesn't yield that return he tries to split the group up. He splits it up by marital status and sex—married couples, single women and single men (single might mean divorced or separated too). Notice that this list of results was derived from the original raw address list by a simple database query of the form: sort by title. The "Mr." and the "Ms." or "Mrs." titles were grouped appropriately or first names were compared to a list of names sorted by gender. Then the database totaled the number of positive responses and the percentages. This could not have done this by hand. The original list was 10,000 and follow-on mailings might be 100,000 or more. A database system must do this procedure and possibly a bar code reader. When he totals the results from these three sub-groupings he gets better results. Results now are: married 2.4%, single women 1.6% and single men 1.1%. Better results, but still just barely past the break-even point, there won't be enough to live on. Married couples are close but still not very profitable as a group. Single men seem like a waste of time. Their number is so low that it seems that he could safely skip consideration of them for all time. Yet that would be a big mistake! Deeper levels of analysis will show the error of doing that.

Next Ronald adds in the type of housing and searches on six separate categories: married people in single family homes or apartments, single women in single family homes or apartments and single men in single family homes or apartments. The results are: married, house 2.9%, married, apartment 1.7%, single women, house 2%, single women, apartment 1.4%, single men, house 4.9% and single men, apartment 0.5%. Look at these results: those single men that he was ready to dismiss out of consideration have a lot of potential. There are two distinct subgroups: single men in houses often have custody or joint custody of children. Single men are notoriously poor shoppers and apparently having kids requires them to buy gifts for their kids and for neighbors kids' birthday parties. These single fathers find it much easier to shop via mail-order then to find the right store and the right age-specific gift in a retail shopping mall. Very interesting, very profitable, this is not obvious from the lower two levels of analysis. Now Ronald could put together a mailing list of groups identifiable from mailing addresses and average more than 2.5%.

He can do better yet. Another variable that effects mail-order purchasing is household income. If he could tell what the household income is for each address then he could fine tune the mailing list to be even more profitable. For individual households this is impossible; only the IRS knows and they won't tell. Using public domain census data to some extent is possible. However, there is a better way for his to get an idea of a household's income. It relies on the fact that home prices and apartment rents are related to household income. (There are many exceptions here; a family that once had a smaller income might stay in their old neighborhood. Still, this is about as good an estimate of average household income in a small area as you can readily obtain). Usually, people will buy as much a house or rent as good an apartment as their income will allow. He can readily find such data in those free real estate catalogues that are available at most grocery stores and drug stores. It will take him some work with a map to figure out which street addresses fall within which housing developments but he can do it.

With the table (we have already seen it in the previous chapter) he can now form a set of rules that allow him a very profitable mailing list. Mail to: 1) married couples in houses in valued between $150 and $200K, 2) single women in houses valued above $200K and 3) single men in houses valued more than $150K.

What is our conclusion? The mass mailings don't take much effort; Ronald's most time consuming task is to process the positive replies. His maximum is about 5000 positive replies a year. He has a large enough franchise-assigned area to allow him to get this many positive replies. Recall that the rate of positive return is the sole parameter that determines income. The higher the rate of positive replies, the less mailing expense that he must incur to get 5000 customers, therefore the more income. Another way of looking at his problem is to notice that the franchise agreement fixes his sales area; so the potential overall mailing list is fixed. They give the sales catalogue to him. He can't vary it. He can change nothing on his part, except what people he chooses to mail it to. That one variable is the only thing that he has control over; he needs to do the best possible with that variable, his entire income depends on reducing mailing costs to attain that 5000 customers that he can process per year. We show the results of several levels of analysis in the following table.

Mail Order Plans Comparison

Mail To:	Sales Rate (%)	Profit ($) with 5000 Customers Needed	Computer Skills/Analysis
all	2.0	0	none
all married list	2.4	16,000	simple observation of
all married and single men in houses	2.7	26,000	database knowledge
income plus related groups	5.1	61,000	database knowledge getting data from CD-ROM

All these methods require about the same amount of work. Recall that most of the actual work is processing the 5000 orders a year. That number of orders will give about $100,000 income no matter how many catalogues he mails out. Mailing expenses is the sole item that he has much control over and reducing that one quantity is the point of his analysis. Big differences are apparent. The no-thought-mail-to-everyone approach is a lot of effort for nothing—literally nothing. Some thought and observation of the incoming orders will lead to a profit of $16,000, too little for him to live on comfortably. By applying a commercial database software system with some knowledge of how to do basic logical queries he can make more than a minimum living with $26,000. Still, the real profit comes with analysis that brings in other relevant data into consideration. That requires more computer and analysis skills, although no more math skills. More computer skills often translate to higher pay or profit and this example is an excellent illustration. Advanced math skills were not needed. Computer software skills with a database system and simple logic were all that he needed. You can do it too. A later chapter will show you how.

Middle School Counselor Case Study

Money isn't everything; sometimes your job involves decisions and you would like to do your job as well as possible. That's where our 6th grade counselor Cynthia is at. We saw her problem in the previous chapter: how to develop a more accurate method of predicting the success of 6th graders in academic level math classes. Before she had each individual student's 5th grade math grade, their 5th grade standardized math test score and historical data involving hundreds of other former students at her middle school. She knows that performance in 6th grade math tracks that of 5th grade math mostly. That means if a child averages 80% in 5th grade math they are less likely to average 90% in 6th grade math and vice versa. Also, she knows that the academic level math is tougher and that the average grade does run a few percent less. Lastly, she knows that a student's standardized test score figures in prominently.

Notice that all of Cynthia's knowledge is in the form of general rules, trends and observations. When she is discussing a student's potential with parents, she can repeat all of that. She can tell the parents of the vast amount of data she has seen and then she asks the parents to trust her judgement. Some do. Some don't. So, some capable girls take academic level math; some don't. The "don't"s are Cynthia's concerns.

She wants a quantitative method to support her recommendations. She isn't going to rely blindly on it. She isn't looking for some formula that takes past grades and test scores and automatically assigns students to classes. Kids have traits that don't show up on grades or tests and Cynthia factors these traits in. She is looking for a good quantitative supporting tool to convince skeptical parents. As start she could simply average all 6th-grade academic level math grades, which is 83% and use that as a quantitative prediction method. That is a gross average with no individual information factored in at all. It wouldn't convince anyone. Another basic rule is that generally students do about the same in 6th grade as 5th. So a quantitative prediction method would be *grade_6=grade_5*. This is somewhat individualized. This formula leaves out many factors and when applied to her data it results in a large standard deviation. More accurate would be a formula like: *grade_6 = grade_5 - 3%.* This takes into account the principal that students do about the same from grade to grade. This formula also considers the additional fact that 6th grade academic level math is tougher and grades will be slightly lower, about 3% on average. Again this formula takes into account individual results and an overall average difference between grades. Nevertheless, something this simplistic will not sway many parents.

The parents that she is trying to convince are parents of talented girls whom Cynthia feels strongly can pass the course. Yet they enroll these talented girls in simpler math classes because they fear that their daughters over achieved in lower grades and will be stressed out by challenging math. That over achieving fear can be countered by referring to high test scores. So test scores will be an important factor and the quantitative predictor should consider them. But how? Prediction equations like *grade_6 = test + 36* don't fit the data well at all well. Grades in 6th grade track grades in 5th grade better than they track the test scores—but the data clearly shows a significant test score influence. How can she include two factors that both influence grades? Regression is the answer. She can use the 5th grade math grade and test score as inputs (causes, independent variables). Then develop a regression equation that fits the 6th grade math grade (effect). This is more complicated then that simple one-variable linear regression, which hand calculators can do. When she does this (details in the later chapter on statistics) she gets the following prediction equation: *grade_6 = grade_5 - 3 + .5(test - 47).* Now this is a useful counseling tool. It fits the data very well. It is simple. It incorporates the two dominant factors, grades and test scores in one easy-to-understand weighted sum. This formula will help concerned parents to see if their children are actually over-achieving and perhaps headed for a drop in grades. Conversely, if a student really is capable in the challenging academic level math then the parents will see a quantitative proof of that too. That is what Cynthia wants. Furthermore, she can compare it with her data and use it to help identify the over-achievers and under-achievers. Those will be the ones whose grades are much higher or lower than their test scores would suggest. Cynthia can devise a numerical cutoff for identifying such students and point it out to their parents as well. All in all it will be one useful tool. Let's compare the results of different levels of analysis on the prediction tool problem.

Middle School Math Grade
Prediction Methods

Level of Model	How Useful?	Math Needed
The average grade is 83% your child should be close.	none	none
Grades in 6th grade are about the same as 5th grade.	some, but not much	none
Grades in 6th grade are about 3% less than 5th grade. Or a formula: $grade\_6 = grade\_5 - 3\%$	some, but again not much	none really, simple averages
$grade\_6 = grade\_5 - 3 + .5(test - 47)$	excellent	regression
$grade\_6 = grade\_5 - 3 + .5(test - 47)$ plus rules: (regression overachievers: $grade\_5 - test > 48$ underachievers: $grade\_5 - test < 30$	superior	statistics plus other analysis

Most administrators in public or government positions know something about statistics. It is a requirement for most college programs in administration, teaching, counseling, psychology, personnel management and others. This is a good example of why this is true. The first three levels of prediction don't really say much. It is not until you get the regression derived formula that you get an adequate prediction capability. Furthermore—and this is important—the use of the model/equation/formula allows you to confirm its validity by checking it logically and checking it statistically by its fit to the data. Statistics are everywhere in public institutions; validity checks are necessary.

An equation or formula is often considered fair and impartial. This takes some explanation. Although, the inputs, here grades and test scores, may have underlying causes that are debatably fair—the equation itself is impersonal. A decision made by reference to equations or formulas is open. It is not like a decision made by a committee behind closed doors. Results are repeatable: substitute the same values into an equation and the same result comes out. People-decisions are not repeatable: people change their minds, get pressured, or change their emphasis. Though an equation is itself impersonal, the use of equations in decidedly personal decisions, like "is a particular child ready for academic level math?" can be very convincing to some people. Parents like Cynthia; she is friendly, knowledgeable and sincere. Still, when she is recommending to parents that they should enroll their daughter in the challenging academic math class they can be skeptical. What if Cynthia is recommending that because of her political beliefs that more girls should take higher math? They don't want their daughters to be policy pawns and have to suffer stress from too challenging a class. Yet show them that equation. Tell them how well it fits the previous classes' grades. Then you remove that doubt.

Statistics play a major role in the social sciences. If the social sciences are your interest, you may not really like math, you may have avoided it in high school and college. You may now lack the background for a good challenging college statistics course. This book isn't a substitute for a course in statistics. Still, you can get a good idea of what prediction and trend analysis are about by reading this book. Then you can see if you want to pursue a college statistics course.

Bakery Inventory Case Study

Maria is that mall bakery owner with that inventory problem: 1) She can only bake during the morning at a nearby restaurant. 2) She doesn't sell donuts baked yesterday—fresh donuts are her trademark. 3) Baking too many donuts wastes money on ingredients, money which could go for much needed additions to her house and vacations with her mother. 4) Baking too few means running out sometimes, alienating customers and even worse risking complaints to the mall management committee which has to renew her lease every year. Predicting tomorrow's sales is her problem. Since she chooses not to keep old donuts and since she can only bake during the morning, she is stuck. She can only sell what she bakes that morning—too few means loss of income and unhappy customers, too many and she wastes money. It's her problem and hers alone.

She has been in business for a year and has a computer record of each days sales and supply-on-hand. Her expenses are not that much. She pays out about $30,000 per year for the donut shop space and equipment. She has some small additional expenses which cancel out additional income like selling coffee with the donuts. Her main expenses are the ingredients and the rent she pays to the nearby restaurant for using their ovens to do her baking. During her first year she was very aware of customer complaints and over baked a great deal. She wasted a lot last year: over $55,000 by her records. Money spent for donuts that she threw away unsold at the end of the day. Her total profit was $45,000 so looking back at that wasted $55,000 causes her a lot of grief. That money is there waiting for her. All she has to do is better estimate the day's sales the night before. But how?

Those computer records are a start. She can see general patterns in them. Yet she needs specific predictions not general patterns. General-patterns-to-specific-predictions is what data analysis is about. There are many methods of doing this; some are simple observations, some involve simple averages or trends, some involve other data sources. Different methods give different predictions. Some are better than others. One thing Maria can do is simply find the most donuts ever sold in one day, then bake that many each day. It is a safe method: she wouldn't ever run out of donuts and leave hungry, unhappy customers. It is safer then the method she used last year, when she baked 2500-3000 a day; very seldom did she run out and it only cost $.15 per wasted donut, but that added up over a year. Every night, however, she will throw out a lot of unsold inventory. Inventory that comes right out of her profit margin. Money that she could use to fix up her house or take that bus tour of Italy with her mother. A better method would be to rely on the pattern that dieters often avoid food early in the week. Since most dieters start Monday, but then get off the diets later in the week. She could get the maximum sold each day of the week for the last year. This is an improvement, again safe but with less waste. A third level would be to consider average daily sales and the most ever sold in one day and sell a set number per day that is somewhere between. A spreadsheet allows her to change this set number. Then calculate what her profit would have been last year if she had baked that many (2000 per day for this simulation) rather than what she actually baked. This is a simple form of a "What-If" study that people use models for (more about this in the spreadsheet chapter).

Another level of analysis would be to consider the weather effect but only on a very large scale. The scale is seasons. More donuts will be sold in the colder weather of winter than in the moderate weather of spring or fall. Even fewer donuts will be sold during the hot days of summer. A spreadsheet will allow her to do seasonal averages and adjust her baking up or down. The highest level of analysis would be to consider the weather day-by-day and to factor in what shopping season it was. She notices a great variance in sales when the weather changes drastically; she feels that the weather is the single biggest factor in her sales after day-of-the-week. Also, she knows that they are more mall customers during peak shopping seasons. However, she is unable to separate out its effect from the day-of-the-week trend and the overall random day-to-day fluctuations in sales. Yet her computer can! She can give all the data to a neural network program and let it predict the next day's sales. Then she can use that as an estimate for how much to bake the next day.

Results of the different prediction methods are listed in the following table. The absolute best possible result would require perfect foresight and is impossible of course. It is included to show how close to perfection the neural network and statistical methods were.

Methods for Predicting Bakery Sales
A Comparison

	Method	Profit ($)	Skills required
	Bake overall Max every day	-35,000	none
	Bake daily max each day	29,000	normal judgment
***	Maria's day-to-day estimate	45,000	experience, observation
	Bake optimal fixed number	52,000	simple spreadsheet skills
	Bake daily max adjusted for season	57,000	intermediate-level spreadsheet
	Multiple Linear Regression	73,000	statistics
	Neural Network prediction	73,000	10-year old running a neural network program
	Neural Network prediction	84,000	Ph.D. running a neural network program
	Absolutely Perfect Foresight	105,000	Conversations with God

Look at the results: the simplistic-no-thought-required method of baking an enormous surplus each day lost money. No business could survive that much waste. A more reasonable plan is to be safe by always baking enough but fitting your baking to the weekly trend—lowest Monday, increasing daily till Saturday. Still, the average daily sales are nowhere the maximum daily sales which occurred on a very cold day. So this method also involves a large amount of waste. The rest of the methods are not absolutely safe. An unusual day might see her running out of donuts to sell. The rest of the methods are designed to have at worst a very small shortfall averaged over a full year (for example Maria's method had a shortfall of less than $5,000 of sales). Maria's method of observation and her experience allowed her to choose a more profitable plan with that $45,000 profit. Still, there are better methods available for her if she uses data analysis. The next two methods could be done with spreadsheet analysis. The third method uses that "What-If" method we discussed earlier. The fourth is making an adjustment for seasonal variations. Maria could do more spreadsheet analysis and perhaps do somewhat better. However, to really optimize she needs a program that accurately reflects the sales variation due to the season, daily temperatures and rain.

A neural network is the simplest and least demanding method for such analysis. It is simple enough for a 10-year-old. As a test, I gave my 10-year-old son Cameron a quick tutorial in how to run a neural network program. Nothing complicated, three hours of instruction in two or three parts (10-year-old boys don't have long attention spans). He learned how to put the data in and set the software running; he essentially ran the program with the default values provided by the software manufacture. Notice that he beat all the rest of the methods, tied the advanced statistics method and came very close to my predictions which also used neural networks. Sometimes neural networks work remarkably well with almost no knowledge required. That is why they so impress me for small to medium sized problems. But more on that in the chapter on neural networks. The moral of this story: there are many levels of analysis. Just because you are using one of the most accurate methods doesn't necessarily mean that you need advanced mathematics. Sometimes just software is enough.

All three of these examples are meant to drive home three important points. There are levels of math and computer expertise. There are more than one way to predict or do data analysis. How well you can do data analysis is directly related to your math skills or the computer techniques you use. The more you know, the more money you make. Blunt and simple! Also, important: if you don't have advanced math knowledge sometimes computer software will save you.

Levels Of Modeling

There are three main approaches to determining a mathematical model of a system. First, the method based on first principles when you derive the equations from other equations based on physical principles. This is the preferred method of physicists, engineers and astronomers. The analog in business is a business process simulation. In recent years these have been popular. However, trained systems engineers who use complicated software to do the task usually develop these models/simulations. For the rest of us, this method is usually difficult to implement since everyday data problems like inventory, sales, production, pricing and others usually cannot be mathematically modeled exactly. Another reason why this method is used much less is because of the mathematical complexity. Often advanced calculus and differential equations are needed to even understand this method. Pros: extremely accurate, Cons: narrow applications, a large amount of math training needed: advanced calculus, systems theory and differential equations.

Second, is the method of regression. It is the most widely used, the most flexible and often is fast to run on a computer. You can apply this method to an almost limitless set of applications including inventory, sales, production, and pricing. The requisite math background is not that much. Except nonlinear regression, for that method an understanding of calculus and a course or more in statistics are needed. However, some regression methods can be done with much less math; we cover these in the statistics chapter. Pros: fast, wide range of applications, Cons: college statistics and calculus courses *sometimes* required.

Third, is a new method in use since the 1980s: neural networks. You can also apply this method to an almost limitless set of applications. Furthermore, it requires less math background from the user than statistical methods. High school math should suffice. The one drawback is that it takes more time to run large scale problems on a computer. The underlying model is close to that of a human brain. Neural networks mimic the human brain in its ability to take large amounts of unstructured data and develop predictive math models from this data. Neural networks work with minimal user intervention. It works much like a child learning something new. However, this "child" has blinding computational speed and once it has learned the data it is much faster than the user. Pros: wide range of applications, minimal math required, Cons: more computer time required.

Commercial Software For Data Analysis

There are many different facets to the general field of data analysis and there are several other areas that overlap. This section tries to clarify the situation for you and to point out the types of commercially available software in each area. We explain and categorize the common buzzwords and software types. Then the book's goals and methods are placed in perspective; their level of difficulty is discussed compared with the other areas. Finally, software is placed in our main theme diagram *data->model->plan.*

Central to data analysis is a model. A model can be a conceptual, logical, mathematical or computer model. Our discussion usually describes mathematical function models or computer models. The term "simulation" or "simulator" is close in concept to a model. Often process simulations are more complex and might include several different models of different sub-processes. Still, programs that develop models are developing simulations in a sense and vice versa. An example of a model is a function that predicts sales of a product in a sales area. Here the model is a function of the average income and population in that sales area. Suppose an analyst uses his knowledge of sales theory to come up with an exact relationship: $sales = .1 * population * (15 + .002 * income)$. That would be an exact or specific math model. This is a process of deriving the model from sales theory without data. This is a completely acceptable method of modeling or simulating a process (much like engineering models). However, it is not the focus of this book. Overall, they can seldom apply this method and it takes a higher level of mathematical background. Now suppose the analyst couldn't come up with the exact model but instead got a more general relation say: $sales = k * population * (b + c * income)$ which captures the general form of the relationship. Now the analyst could determine the unknown coefficients k, b and c by fitting data to this general model with regression. This is widely done; perhaps it is the most common form of developing models. This *general model form + data -> model* is discussed in this book since assigning a general form doesn't change our methods. So this common form is part of the theme of the book. The last form is where the analyst knows nothing about the general form of the sales relationship. Or when the analyst chooses not to assign a general model form but lets the computer choose it for him. This is also a theme of this book. It is also a very common situation. Here the general form that the computer program (a neural network or some data mining program) assigns might differ considerably from the general form we saw above. There could be different types of functions, maybe trig functions or exponential functions. Still, the resulting model will fit the data. We will capture both of the data modeling methods (with and without a general model) in our theme diagram data->model->plan.

The following diagram lists the most common types of software and attempts to place each within our diagram. (Caveat: there aren't fixed industry naming conventions so you might see a software product advertised that includes capabilities outside this diagram. However, this diagram does help you place different software capabilities.)

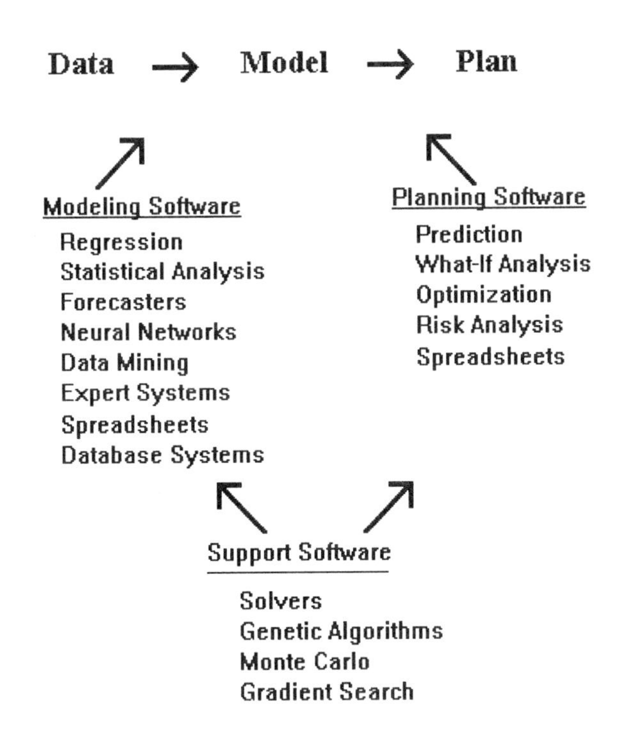

This diagram shows common software systems or techniques used to turn data into a model and then a plan. This model might be a math function, a computer program or a set of rules. Some of the techniques require more math to understand. This book's line of demarcation is statistics: we will concentrate on those techniques that don't require a course in statistics (this usually rules out calculus too). This book's plan is to focus on the easier data->model techniques: spreadsheets, database systems, regression, neural networks and some study of expert systems. Forecasters and data mining tools often employ statistics, regression and perhaps neural networks as part of their techniques. If you run into software of this type, you may understand how to use them by reading this book.

Support techniques often come as add-ons to spreadsheets. Their difficulty is harder to gauge—often you can run them with default values, so you don't need knowledge of advanced math. However, if you really want to understand them, you will need calculus, statistics and more. These are the mathematical routines. Gradient search is an advanced calculus technique that is widely used in solver programs. Genetic algorithms are hot buzzwords these days. These algorithms try to optimize functions based on algebraic operations that mimic evolutionary DNA processes. The analogy being that "better" means farther along the evolutionary chain in evolution theory and "better fit" means minimizing a function in data fitting theory. So the goal of evolution of the unknown coefficients (the analog of DNA) is set up to be minimizing a function. It may sound strange—but it works. Monte Carlo techniques are also very useful. In effect, they try many random variations and pick the best (random as in gambling, hence the reference to Monte Carlo, the island of casinos).

We will not discuss the model->plan software techniques as much as prediction techniques in this book. This is because their usage depends on your particular application (i.e., are you trying to predict trends, gauge the possible effects of an alternative scenario or assessing risk). We will discuss prediction a great deal because that comes right out of regression and neural network techniques. Spreadsheet techniques will be touched upon, as will What-If analysis.

Chapter 4

Math Anxiety
This Is All Nice, But I Could Never Do It!

Did you ever walk into a math test in high school, knowing that you weren't really prepared? What did you feel like? Did you ever spend hours doing a homework assignment in algebra that involved word problems and fractions then go to bed totally frustrated by your lack of progress? You aren't alone. Math anxiety is common in American students. Many fear having to learn new mathematical techniques because they fear doing poorly at it or getting hopelessly lost and frustrated. Data analysis requires some math. Therefore, some people might also dread learning data analysis techniques. This chapter is not about math techniques, nor does it discuss data analysis, it doesn't explain any software, nor offer any technical advice. What it aims at are the twin problems of anxiety and attitudes. If you feel inhibited by math or software, this chapter tries to help you how to break through your inhibition. Math, computers and endless data are facts of life. You can ignore them only to the detriment of your own career and happiness.

Q: I never have math anxiety; I have taken several advanced math courses, done quite well in them and my coworkers respect my abilities and come to me for statistics advice. Finally, I am not the least bit worried about learning new data analysis techniques on my own. What's in this for me?

A: Skip this chapter.

Q: I didn't hate math in school but I have a healthy respect for how difficult it is. I can't really identify why, but I have some unpleasant memories. Is that anxiety?

A: Anxiety usually manifests itself as a feeling that you will never understand this new math topic. It is a feeling of a complete mental block, a dead-end. The unpleasantness associated with mathematics is largely due to a few characteristics of elementary through high school math training. A few things cause a lot of math anxiety. 1) Word problems—many people dislike or fell inadequate with word problems. 2) Tests—most people get anxious about tests, especially math tests. 3) Fractions—these cause a lot of anxiety. 4) Math is a different language—because of this it often seems abstract and difficult.

Q: Yes, your list seems right on the money. I remember word problems, tests and fractions as my biggest complaints too. You can't hide the fact that data analysis includes some math. So, why won't I hate data analysis too?

A: Data analysis is different. 1) Word problem ability is only needed to a small degree. Any data analysis problem you are working on will (presumably) be in an area that you understand well. This will make this issue easier to deal with than it was in high school algebra, where the word problems were in areas that you weren't familiar with. 2) Tests—I am *not* giving any tests in this book—case closed. 3) Fractions—these don't appear in the examples in this book and you, the data analyst, don't personally have to solve fraction problems anyway. Fear of fractions shouldn't keep you away from the kinds of data analysis discussed in this book. 4) Math is a different language—this one does apply to data analysis since we are stating business problems in terms of equations and logic. However, the business area should be familiar to you.

Q: Math anxiety—I have it bad! I can't understand word problems, I can't remember formulas, I can't even remember the stuff from last semester's math class. What chance do I have?

A: Math anxiety is common, very common. Look, math isn't easy, no one says it is. However, it is understandable by most people. Math anxiety can strike anyone, even the mathematically gifted. The feeling is the same though the level of math may differ. Anxiety strikes third graders who are frustrated to tears by multiplication tables. Anxiety strikes high school students who can't cope with algebra: letters where numbers should be. Anxiety strikes college students who can't make sense of calculus concepts though they are repeating the course for the third time. Finally, anxiety strikes graduate math students, who everyone considers gifted. When research problems stump them and they wonder if are as good as their peers. You are not alone.

Q: Math anxiety isn't my problem. I really am not good at math and I don't feel that attempting higher level math holds any hope. I'll probably just fail, won't I?

A: Not so fast. Just how "not good" are you anyway? Studies have shown that average college students have all the mental capability needed to pass statistics and calculus. Now it may be that an "average" student will spend significantly more study time. It may be that an "average" student probably will feel more anxiety and that an "average" student might just get a "C" in the course. Nevertheless, an "average" student can pass.

Q: I really want to get a Master's in educational administration. I've always dreamed of being a middle school principal, I always knew that I could make a difference in kids' lives at that critical time. However, that program requires a course in statistics for research methods. In the past, I have been anxious about math and statistics could cause a lot more. I am thinking that I could skip that anxiety: the stress, the insecurity and the uncertainty by switching to another graduate program.

A: Don't do that! Register for that statistics course and follow your dreams. First, statistics is not that hard as we have already seen, the average student can pass it. Second, living involves some stress, some anxiety, some risk taking. This type of stress might be avoidable by skipping statistics, but look at the price. You miss out on a chance at a long running dream. You may have to study extra; you may have bad days where you question your abilities; you may only get a "C." But almost everyone passes and even if you fail you can retake the course. In the big picture of life this is not bad. This is a situation where you have to live with your decision to take an easier path. You may regret your decision years from now. My favorite example of this quandary is a speech that Mel Gibson gave to his troops in the movie *Braveheart*. Mel Gibson often plays a slightly mad character and his role here (William Wallace, 13th century Scottish hero) was no exception. Trying to convince an outnumbered, lightly armored band of Scottish farmers to attack a professional English army instead of fleeing, Mel (as Wallace) said " . . . Fight and you may die, run and you'll live, at least a while. And dying in your beds many years from now, would you be willing to trade all the days from this day to that for one chance, just one chance, to come back here." Many of his troops did trade away all their tomorrows that day—the ground was covered with their dead. But they won the battle. Aggressive, confident people often do. I don't recommend rashness; however, here we are not talking about armored soldiers, a hail of arrows and sudden death. This is just math. Years from now don't lie in your bed regretting not registering for that statistics course. If you need that statistics course to follow your dream: be a little like Mel. Years from now you will have no regrets.

Q: Wow! Armored soldiers, a hail of arrows, and battlefield speeches too. Sends shivers down my spine. But I am one of your women readers and as cute as Mel Gibson is, this storm-the-gates attitude won't work for me. This anxiety stuff is real. I still ask: "I am thinking that I could skip the stress, the insecurity and the uncertainty by switching to another graduate program."

A: OK, different approaches work for different people. And that 13th century, sword wielding Scotsman probably never took math anyway. Here is a modern day expert on math anxiety: Sheila Tobias, author of the book *Overcoming Math Anxiety*. Sheila is speaking of math anxiety when she says: "It is important to go through this at least once, because managing anxiety is just that: experiencing anxiety and mastering it." The point is that reading about math anxiety doesn't get you through it. Studying math when you are anxious about it and then learning math in spite of anxiety teaches you how to handle math anxiety. Don't avoid math that you need because of some possible anxiety.

Q: Would women treat this situation any differently?

A: Many would and that is part of Sheila Tobias' theme. She points out that studies show that women often attribute math failure to lack of ability; men often attribute math failure to lack of effort. There is a clear and prominent difference between the attitude that: "I was not smart enough to do it" as opposed to the attitude that "I didn't spend enough time studying." Men fail math too, maybe more often than women. Yet among the group that has enough ability to pass, more often men will simply jump in and take the course. Also, Sheila argues that math anxiety may not be distributed equally between the sexes. She says also: "... I still argue that excessive anxiety inhibits women more than it does men." In essence, it is believed that some women create an extra hurdle for themselves by worrying more about math than an equally talented man would. Knowing this might help you get rid of that very same anxiety.

Q: Speaking of women, men and math, just what is that math "gender gap" that the media talks about? I know that high school males do better on standardized tests like the SAT, but how big and significant a difference are we really talking about? And does it last past high school?

A: This is one of the biggest issues in education today. For some reason(s) male students at certain levels do better than females on standarized math tests. At the elementary school level there is no difference, but scores start to diverge in middle and high school. There is about a 40-point difference in the SAT Math test. This is just an average and by itself doesn't mean much. In addition, one intriguing fact about this whole business is that: studies by several colleges show females doing as well on advanced math courses as males even while being outscored on the SAT. In other words while better SAT scores do correlate to better math performance in advanced courses the scales are shifted between the sexes. For example the average male student that gets a B+ in an advanced college math course might average 620 on the Math SAT; the average female with a B+ might average 590. All of this should give you pause if you are considering avoiding math courses. Still, there is a problem here. The problem is that this 40-point average translates to a shift in the normal distribution of male and female scores. So at the highest levels (like SAT Math of 750) males outnumber females by ratios of 3 to 1 up to 7 to 1. Since the highest scores are most visible to the media, they pay a lot of attention to this gender gap. In addition, universities give the most highly paid college degrees in the most technical therefore most mathematical majors like electronic engineering, computer science and chemical engineering. Now only the people with the best mathematical preparation and test scores major in such areas. Therefore, this mismatched ratio between the sexes is mirrored in the number of students in these highly paid areas. Eventually, it is also mirrored by the number of female vs. male graduates in these technical areas. This leads to newspaper headlines like: "New Engineering Hires Favor Men by 5 to 1."

Q: This looks like discrimination to me. Who is to blame?

A: Corporations say "not us." Our engineering jobs require a new employee to have an engineering degree. The male-to-female hiring ratio reflects the number of engineering graduates out there. Ask the colleges. Colleges say "not us." Many studies have shown the ability to pass math intensive technical courses is closely correlated to standardized test scores like the SAT. They administer the SAT before college. Ask the high schools. High schools say "not us." High school females actually get better math grades than males (that is a fact). The problem is that by this stage females are already scoring lower on some standardized math tests than males. The main problem occurs before now. Ask the elementary schools. Elementary schools say "not us." In elementary schools girls do much better than boys at every subject including math and the test scores are the same. Socialization is the issue; boys come into school more socialized to take math and science. This socialization occurs outside school and before the child even enters school. Ask the parents or blame society.

Q: Maybe my question should not be *who* but *what* is to blame. What factors cause this gap? And can I fix the problems in my own situation?

A: Tough question, there is no known set of factors that completely cause this gap or even are the dominant factors. Worse, if you are more than, say 11 years old, you probably can't change them. Still, you are OK. Maybe you can't change the underlying factors yourself, but you can compensate for them even late in life by studying the right kind of math and data analysis. Anyway, you might be past college by now. You might simply be wondering how to fill the gap in your math education so that you can analyze the data in your own business area. That is why I wrote this book: not for middle school kids but for adults that needed to fill gaps in their education so they can analyze their own data.

Chapter 5

Statistics

What You Need To Know Without Theory

Statistics—it's not just for liars anymore. In times past it seemed that statistics was mainly used to promote various political views. They have always told us that old Mark Twain quote: "... there are liars, damn liars and statisticians." Most adults' exposure to statistics is the unending barrage of numbers, averages, correlated factors, bar charts and graphs that we see in newspapers or political ads. It often seems that statistics is a theoretical mathematical field whose sole purpose is to give ammunition to politicians, lawyers and researchers. Don't get me wrong; they still misuse statistics that way. However, in the past ten years user-friendly software has brought statistical prediction within the reach of the casual user of PCs and hand calculators. Great problems: federal government policies, global predictions and the like are routinely done with statistical methods. Smoking and lung disease have been subjected to so much publicized statistical analysis that one commentator (Fletcher Knebel) joked the reverse was true also: "It is now proved beyond doubt that smoking is one of the leading causes of statistics." Still, those of us with more mundane problems can make good use of statistics. Statistics is a prediction tool. In this chapter we look at three main topics in statistics: correlation, regression analysis (trend prediction) and that most infamous of all statistical curves, the normal or bell curve.

This chapter deals with statistics for data prediction; this is *not* a condensed course in statistics. Most topics in a college statistics course won't even be mentioned. For example: Chi square, confidence intervals, exponential distributions, means, modes, percentiles, ranks, skewedness and variance are just a few of the topics that we will not cover. And just how can we do this? Well, statistics has many uses only one of which is prediction. However, many of the topics in a first course in statistics are aimed at other purposes. Designing experiments, collecting experimental data, measuring features of data like variance or modes or means take up much of a college statistics course. All of this is necessary if you are going to be a psychologist, health care manager, government auditor, or quality control specialist. However, if you want to find trends in your data and predict future needs and directions of your business then regression and correlation analysis might be enough. This chapter attempts to bypass a lot of the theory and definitions. The hope is to get you directly into real world examples of statistical prediction and trend analysis of data sets.

Correlation

Newspapers are full of items that state facts like the following:

- Lung cancer rates are closely related to the number of cigarettes that you smoke.
- A new study shows a relationship between violence on TV and teen crime rates.
- A survey found that students from lower income areas receive lower grades on the average.

Implicit in each of these statements is a (potential) cause-effect relationship. Smoking causes lung cancer (well established), TV violence causes teens to commit more crimes (debatable), or low income causes low grades (some effect but there are several others).

Cause and effect, effect and cause—this relationship is the cornerstone of most scientific theories about almost anything. Sunspots and cancer, education and success, low inflation and low mortgage rates, voter discontent and election turnout are relationships that are known cause-effect relationships. Functions are usually used to describe cause-effect relationships between measurable factors. In a sense the cause is the independent variable; the effect is the dependant variable or function value. If "f" is a function that describes a cause-effect relationship then we expect that *f(cause) = effect*. This situation is ideal. We have completely modeled the phenomena of interest by this functional description. If we want to gauge the effect of an action, all we have to do is substitute the value of the cause into "f" and compute the effect.

Ideal as it is to have a functional relationship between two factors, often it is not necessary or possible. Maybe we want to identify factors related to a phenomena that we observe and that is all. Nevertheless, often, far too often, we don't know all the factors that will cause an effect. Maybe we know one. Or maybe we think that we know one. So we need a measure of whether or not a factor is a cause of another factor.

The correlation function is a way of studying possible cause-effect relationships between data. Correlation analysis is weaker than logic. This is because correlated variables need not be related in a logical cause-effect relationship. Suppose correlation analysis states that factors A and B are correlated. What does this mean? It can mean one of five different things: i) factor A is a (partial) cause of factor B; ii) factor B is a (partial) cause of factor A; iii) factors A and B have no relation to each other; iv) factors A and B are effects of the same cause(s) but neither causes the other; or finally v) factors A and B have a complex interrelationship where both are causes and effects of each other. These represent several possibilities. Examples will help. A classic book on misusing statistics is "How to Lie with Statistics" by Darrell Huff; in it he gives two examples. First example case iii): "There is a close relationship between the salaries of Presbyterian ministers in Massachusetts and the price of rum in Havana." Indeed that was case although there is no logical connection between the two variables. This example shows how simply having two correlated variables can sometimes mean nothing at all. Second example, case v): "The more money you make, the more stock you buy, and the more stock you buy, the more income you get; it is not accurate to say simply that one has produced the other." Here stock purchases are indeed correlated to higher income. However, which comes first, more income or stock purchases is not clear. "Chicken and Egg" arguments arise here; if two factors are correlated which came first, which is the cause? Knowledge about the process or phenomena is necessary to answer this type of question. Mathematics alone can't do it. Correlated variables or measurements are very common in all social sciences. Smoking is correlated to lung cancer, divorce is correlated to poverty, education level is correlated to income, amount of study is correlated to grades, and on and on.

Exactly What is Correlation?

Correlation is not some magical universal relationship between phenomena that transcends different sciences; it is simply a mathematical measure whose value varies between -1 and +1. Phenomena whose measurements give values above about .3 (or below -.3) are said to be correlated, those above about .7 (or below -.7) are said to be highly correlated. Correlation is calculated as a number that measures a degree of relationship between two parameters measured several times. (Like for a population of people two numbers each: estimated number of cigarettes smoked in a lifetime and length of life, or for a population of high school students two numbers each: family income and grade average.) Correlation is based on a very simple yet powerful idea. When two parameters (say X and Y) are related to each other the following *usually* happens. An above average value for X means an above (or below) average value for Y. For example: smoking twice as many cigarettes leads to a much higher than normal chance of lung cancer then just smoking the average number. Smoking only half the average number leads to a still lower chance of lung cancer (of course not smoking leads to the smallest chance). Both facts are part of the relationship between smoking and lung cancer. Correlation is a measure or index of how tight this relationship is. To understand correlation first consider the two variables X and Y and their respective average values or means X' and Y'. For each corresponding pair of values X and Y form this product: (X-X')(Y-Y') which is the product of how far off from average each measurement is.

Correlation is based on this one simple, powerful observation. The sum of this product over both data sets tends to a *large number* (positive or negative) for *variables* that are *related.* Conversely, this sum tends to be *close to zero* for *variables* that are *not related.* Here is why. For definiteness let's assume X higher than average means usually Y higher than average and X lower than average means usually Y lower than average. (The opposite case will be discussed next). Now, if X is above average then Y is usually above average so both terms in the product (X-X')(Y-Y') are usually positive. Alternatively, if X is below average then Y is usually below average so both terms in the product are negative. In either case, the product is positive. Therefore, a sum of many of these products will be a large positive number. The second case is an inverse relationship when X higher than average means Y lower than average and vice versa. In both of these cases the product is negative and therefore a sum of these products will be a large negative number. So what? How can this tell anything? Well, the point can be seen by looking at the case where X and Y are not related at all. Now if X is higher than average then half the time Y is lower than average and half the time it is higher. This leads to roughly an equal number of positive and negative products of roughly the same magnitude. Therefore, a sum of many of these products will involve a lot of cancellation of positive and negative numbers and will be close to zero. Therefore, related variables X and Y form a sum Σ (X-X')(Y-Y') that is a large positive or negative number. Unrelated X and Y form a sum that is close to zero. The actual correlation function scales the sums and products so that a perfect relationship gives either a correlation of +1 or -1. The point of all this: if you don't know if X and Y are related data sets then you can compute this sum of products and that will tell you. That is all it takes: a simple summation of a lot of multiplications and you have a measure of cause-effect. And that is the idea behind correlation.

Two parameters that agree in value a high percentage of the time are *not* necessarily correlated at all. For example: ancient priests kept ignorant masses at bay by holding elaborate ceremonies to beseech the gods to let the sun rise each morning. If you didn't know any better and every day of your life there was a ceremony and every day the sun rose, then you might accept a cause-effect relationship here. A second example: the University of Nebraska football team wins 85% of its games. It doesn't rain in the Sahara desert more than 99% of the time. On 84% of the Autumn Saturdays it doesn't rain in the Sahara and

Nebraska wins. Do we have a cause-effect relationship here? Not hardly, the correlation between these two events is about zero. These are excellent illustrations of why correlation can pick out potential cause-effect relationships when straight percentages fail.

How Can You Use Correlation?

Correlation is used everywhere, in all details of everyday life. Number of cigarettes smoked : length of life (- correlation), stock market average : weather in Montana (0 correlation), attendance at football games : amount of rain (-correlation), height: weight (+ correlation), height : hair color (0 correlation), time spent watching TV : time spent studying (- correlation), tossed coin A : tossed coin B (0 correlation). It is all about cause-effect; X causes Y implies X is correlated to Y (usually). (This caveat is necessary because some cause-effect relationships do not cause a correlation). So conversely, X correlated to Y means there is possibly a cause-effect relationship between X and Y.

You can use correlation to determine cause-effect relationships in data; then use the causes as inputs and the effects as outputs to your model program. The main outputs of modeling are functions like *f(x) = y* or *f(cause) = effect* or a rule like: *If cause equals value x then the effect will equal y* or a neural network with the inputs being the causes and the outputs being the effects. Before you can fit the data to a model or a neural network, you need to determine exactly which data sets should be the inputs or causes. We glossed over that earlier, treating the determination of inputs and outputs as given, something easy to do. That is not so in the real world. Often we are faced with many different data sets and only a few are relevant to our problem. Reconsider the baking problem: there are many data sets that have various correlations to the output/cause data set: daily sales of donuts. In this example these were the correlations between the daily sales and other available data sets: sales on this day one week ago (.78), yesterday's sales (.5), unemployment rate (0), Dow Jones Average (0), day of the week (.65), temperature (-.4), rain (-.25), number of mall customers (.85), and sunspot activity (0). You can quickly reject national economic and astronomical factors as having too little influence on sales. Two moderately correlated factors need consideration: sales on this day last week (.78) and yesterday's sales (.5). Should you use them as inputs? These are sales numbers like the output quantity that you are trying to predict. Clearly, the same factors that effect tomorrow's sales control them. But they are not independent factors; sales-on-the-same-day-last-week depends on the day of the week and the weather that day. Since Maria, the store owner, doesn't see any week-to-week trend the answer is probably: no. Just use the independent factors: day-of-the-week, temperature, rainfall and season. (However: this is always the case in stock market predictions. There previous data of the same type is routinely used, since market confidence ebbs and flows from week to week.) How about that number-of-mall-customers (.85) variable? It looks great. Yes, it does, more customers, more sales, fewer customers, fewer sales-it makes sense. Still, it won't be available until too late. You won't know how many customers visited the mall until the day after—too late. An astute reader might note that mall attendance also depends on the day of the week and the weather; so actually donut sales and mall attendance are both effects of the same set of causes, day of the week, temperature, rainfall and season. One final but important point: the independent variables didn't necessarily have the highest correlation with sales (day was .65, temperature was -.4, rain was -.25 and season was .43) but they formed a logically independent set of variables that each logically contributed to the sales.

Too much data and too many data sets to choose from is often a problem. Certainly, it's true today when you have more data available from the Internet then you can possibly analyze. But is too much data really a problem? Suppose you just use all of the data that is correlated to your predicted output variable; you just input all of the data to our regression programs or neural networks. Can't the programs just ignore the other data and then you won't have to make the decisions yourself? Unfortunately, no—you still have to think for yourself. For deep technical reasons using highly correlated data sets as inputs to neural networks or regression programs causes major problems. (Aside: these deep technical reasons have to do with the program's search routines getting stuck in long narrow valleys in the n-dimensional topology that represents the state space of the unknown variables. Or the program's matrix inversion routines having trouble with singular or nearly singular matrices—which is much more then you want to know.) Independent input variables work best. Day-of-the-week, temperature and rainfall are independent data sets, there isn't any correlation between them. This is the kind of analysis that you will go through prior to building a math model or a neural network. All situations are different and you must analyze your own data sets for appropriate input variables. Correlation is a good tool for this.

Nor do you have to be an MBA, business analyst or entrepreneur to need correlation analysis. It can improve your understanding of business practices even at the level of administrative assistants or personnel officers. For instance, take some normal data sets available to personnel officers in a company: for every employee: grade/salary, amount of training taken each year, times of training, times of vacations, job descriptions (administration, engineering, managers, sales), trip times and others. By doing correlations between pairs of these data sets you might discover relationships or tendencies like the following. Amount of training correlates strongly to salary (maybe you could suggest more training for the lower paid personnel). Or timing of training correlates to vacation timing (this would deplete your workforce too much during the summer—maybe suggest to your boss that she emphasis training more often in the fall or spring) or other relationships. A later chapter, *Lost in a Large*

Corporation, goes into detail about this. Correlation is not just number crunching to make charts and table; correlation helps you understand your job better.

How to Calculate Correlation

Up to now you have seen what correlation is and how you can use it. You have your data sets and you are ready to go, but how do you actually calculate correlation? The simplest way is to pick up your $20 calculator (only very inexpensive calculators don't have some correlation and regression capability). The one I am using now is a Sharp Scientific but I also have a Texas Instruments Financial calculator and a Casio that calculate correlation about the same way. Let's try a simple example where both the X and Y data sets are all 1s or 0s.

X 0 0 1 1 1 0 1 0 0 1 Y 1 0 1 1 1 0 0 0 0 1

Calculating Correlation

Step

1 Turn the calculator on and put it in statistical mode. (I have to key in "Mode", then a "3" then a "1", but yours will be different. Your calculator will probably be different for the rest of these steps too, but the basic idea will be the same.)

2 Key in the first X value, 0 in this case, then enter it. (I press a key labeled "(x,y)" again yours will be different.) Key in the first Y value, 1 in this case, then enter it. (I press a key labeled "Data"; again you may differ).

3 Key in the second X value, 0 in this case, then enter it. Key in the second Y value, 0 in this case, then enter it. Continue keying and entering all X and Y pairs until you have entered all ten numbers in both data sets.

4 Recall the correlation value. (I press "RCL" then "r", yours will differ) and the value .6 is displayed on the screen. You are done!

At most you have to key in twenty numbers and press five or six other keys. You can do it all in a couple of minutes. It is that easy. An exercise that you can try is to calculate the correlation between data sets A and B below. Comparing the correlation values between these two examples is very instructive. Consider the following four sequences of numbers which represent events that either happen (1) or don't happen (0): you have already calculated the correlation (.6) for the second pair, now calculate the correlation between the A and B sequences, you should get -.1 essentially no correlation at all.

Correlation doesn't simply measure agreement between measured entities. It

A 1 1 1 1 1 0 1 1 1 1	X 0 0 1 1 1 0 1 0 0 1
B 1 0 1 1 1 1 1 1 1 1	Y 1 0 1 1 1 0 0 0 0 1
80% agreement -.1 correlation	80% agreement .6 correlation

works in a more subtle way than that. The A-B pair for example: suppose the two events were linked in a cause-effect relationship. Then when the rare occurrence of a 0 appears in either one we would expect a 0 in the other. Not in the A-B example though, and that is precisely what the correlation function picks out. The X-Y example shows both sequences where 1s and 0s are both equally likely (probability = 1/2) so one would expect agreement between X and Y about half the time. However, the actual agreement is 80% so there must be some relationship. Conclusion: there is a possible strong cause-effect relationship between X and Y; there is likely no cause-effect relationship between A and B at all.

Spreadsheets do correlations easily too. The same example in MS Excel is easy to do. I entered the X data in cells A1 to J1, the Y data in cells A2 to J2. Then I pressed the "Function Wizard" highlighted "Statistical" in the "Function Category" and "CORREL" in the "Function Name" then click on the "Next" button. At this point in the window were two places to enter arrays; in "array1" I entered A1:J1 (designating the A data) and in "array2" I entered A2:J2. The correlation value .6 appears in the "value" space at the top of the window. Quite easily done!

Correlation analysis of many potentially useful data sets helps you pick out the most useful ones for your own data analysis and prediction efforts. Don't shy away from it. Correlation is easy to compute with inexpensive calculators or standard PC spreadsheet programs. Anyone can do it.

The Multiplication Principle

The Multiplication Principle is a useful tool to predict the overall probability of an event if the event depends on several factors. These factors must be statistically independent from each other; in practice this is often hard to verify, but the principle is useful as an approximation anyway.

Trait	Probability
Family history	
parents not separated/divorced	1/2
not an only child	2/3
never married	1/2
Physical appearance	
6 feet tall or taller	1/4
handsome	1/5
Lifestyle	
doesn't smoke	2/3
doesn't own a gun	2/3
Catholic	3/10
wants kids	4/5
Financial	
not in debt	3/4
college graduate	1/5

It is easy to understand and to apply. Instead of jumping into probability theory we will look at one whimsical example. Many single people have a mental picture of whom they would like to marry. If you are over 21 you realize that the perfect dream-person might have some flaws if you actually met her (or him); still speculating is fun. Krista just graduated from college with a B. A. but no engagement ring and no steady guy. Her mom and friends have a lot of advice for her on men and marriage, but Krista is skeptical. Their advice seems well founded; all the characteristics that they are telling her to look for and to avoid in men seems reasonable–but there are so many. Will any guy fit all these conditions? Her mom emphasizes family history: she should not marry a divorced man or one whose parents were divorced and she should avoid only children too, they are always spoiled. Her friends emphasize his lifestyle: he shouldn't be a smoker or own a gun (violent tendencies), he should want kids and be of the same religion. Also, Krista should get a college graduate who doesn't spend himself into debt for good future prospects. Krista, who is tall and beautiful, has some desires of her own: he should be over six feet tall, good looking, he should be older (say 24) but not more than five years older. Krista doesn't think that any of her personal desires are unreasonable nor does she think any of her mom's or friends' ideas are unreasonable either. It's the composite guy that might be a problem, with all these conditions he might be hard to find. Being of a mathematical mind, Krista estimates the probability of each of these traits and makes up a table. After that she applies the mathematical method known as the "multiplication principle" to decide how many men really fit this category.

The multiplication principle consists of multiplying all these probabilities together to get one probability for the composite man. One restriction is that the characteristics are independent of each other and for the most part they are. Multiplying all of the probabilities gives:

$$1/2 \times 2/3 \times 1/2 \times 1/4 \times 1/5 \times 2/3 \times 2/3 \times 3/10 \times 4/5 \times 3/4 \times 1/5 = 288/2{,}160{,}000 = .013\%$$

There are about 1.3 million guys age 24 in this country, .013% of that number is about 173 men with Krista's list of characteristics. Even if she is willing to go up to five years older, that still leaves a very small number of men to choose from. In fact, about 17 men per state qualify under those conditions. Figure in the chance of meeting such a guy, assume that he is not interested in someone else and assume that he would fall for Krista. Then Krista's chances of meeting, then falling in love with and eventually marrying a man with all of the traits is very small.

Men face the same problem too, at least they should. A personal observation is that about one half of all men from 18-30 years old (one fourth of men over 30 also) are looking for a beautiful 18-year-old girl with no other necessary traits.

How all of this works is a mystery, some things are beyond the power of mathematics.

Regression Analysis

All sciences have the need to predict. Physics has it easier than most since by the laws of physics, all the causes are measurable and the predictions are accurate. Not so for the social sciences: economics, sociology, psychology, and education all have many unknown or unmeasured factors that influence their phenomena and spoil the accuracy of their predictions. Everyone is looking for formulas or math functions that do things like the following: f(cause) = effect or f(number of cigarettes smoked) = years life span shortened or f(5th grade math grade, test score) = 6th grade math grade or f(day_of_the_week, temperature, rain,shopping_season) = predicted donut sales.

Unfortunately, none of these equations is exactly known in contrast to the situation in physics where exact equations like: $e = mc^2$ abound. One problem is that many other factors are causes for the same effect which don't appear in the equation. (For example, life spans can be shortened by factors such as heredity, sex, poverty to name a few—not just number of cigarettes smoked). Therefore, we can't find exact equations. There is still hope however: the next best thing to an exact equation is an inexact but *close approximation equation*. This is the point of the regression function and regression analysis. What you get is a function f that relates number of cigarettes smoked to *average* years life span shortened. For example: an exact function might say that if you smoked 20,000 cigarettes over a lifetime then you would live *exactly* three months less than you would have. Whereas a regression function would say that you *probably* have shortened your life by three months, but maybe by five months or maybe not shortened it at all. There is a randomness to regression models. They are not exact but still very useful.

How Can You Use Regression?

You can't read a newspaper or magazine without encountering a numerical prediction of something. "Unemployment will be down 2% this fall" or "Family income will increase 3% next year." Both sound good, still economic predictions are always a little distant and notoriously inaccurate. Maybe a more personal headline like "Every additional year of education is worth a $4,800 pay raise." That hits closer to home. That makes all those years of school seem worthwhile, but how did they get such a number anyway? Or the apocalyptic predictions: "Ten million more children will be in poverty by the year 2000" and "Global temperatures to increase 2.1^{o} by 2015 due to the Greenhouse effect." These two foretell of a grim future and are leads to articles that offer views of how to avoid these futures. Yet, where did they get such precise numbers as "ten million children" and "2.1^{o} by 2015?"

Behind these predictions is the science of trend analysis, which is usually done by statistical regression. It is the same as estimating donut sales for tomorrow, or predicting an incoming middle school student's chance of success at academic level math. In those cases, we saw that the bakery owner and the middle school counselor both wanted prediction tools because they intended to do repeated predictions. However, if your goal is only a one time prediction (e.g., the global temperature in 2015) you still need to develop a predictive model. Behind all of those headlines are statisticians that have developed statistical models for those one-time predictions. You have seen the results of regression analysis. We have discussed it in other settings, now it is time to study it in more detail. You need to see exactly what knowledge is required to do regression modeling, what software can do it, and what pitfalls you can fall into with statistical regression analysis. You will see that while the underlying math theory may be tough (it requires calculus) you don't need the theory to apply regression techniques. Your calculators and spreadsheet programs can calculate predictive regression equations from data that you supply—and do it easily. However, trend analysis can be misapplied, often it can be flat wrong. We will look into this issue too.

The goal of trend analysis is a prediction model. Although what you may really want is a number, that number will come from the model. You have to develop a model before you can calculate that single number. It looks like this: *data -> regression technique -> model -> numerical prediction.* The model will be a mathematical equation. Simple enough so far. The tricky part is the form of the mathematical equation. It can be a line, called linear regression, the simplest and most common form, it can be a quadratic or cubic polynomial; it can involve trig functions (called a Fourier series) or it can involve complicated composite functions (called nonlinear regression, this application requires calculus and is harder to apply). There are still other possibilities. The number of different forms of math models is truly infinite. You, the user, need to assign the general form of the model; your computer will calculate specific coefficients. For example: you the user will attempt to model a data set with a quadratic model of the form $f(x) = a + bx + cx^2$ where a, b, and c are unknown coefficients. Your regression software will then fit this general form to your data and return the values of the coefficients (for instance a=.05, b=1.1 and c=.7). The regression model would then be the equation $f(x) = .05 + 1.1x +.7x^2$.

How to Calculate Regression Equations

Calculating regression coefficients for simple equation forms is easy—actually very easy. Simple equation forms (linear and quadratic equations) are by far the most common regression models. Let's start with a practical everyday situation in the business world. Your company recently introduced a new product and it has been selling quite well for the first few months. Stockholders (a greedy lot) are pressing management for projected sales in the future. Your monthly sales data are listed in the following table.

Time (months)	Sales (units)
1	3,000
2	11,000
3	17,000
4	21,000
5	23,000

You pick up your calculator and do a regression fit to this data in a very similar way to the method we used for calculating correlation.

Calculating a Regression Model

(almost identical to calculating correlation)

Step

1 Turn the calculator on and put it in statistical mode. (I have to key in "Mode", then a "3" then a "1", but yours will be different. Your calculator will probably be different for the rest of these steps too, but the basic idea will be the same.)

2 Key in the first X value (time in months from the first column), 1 in this case, then enter it. (I press a key labeled "(x,y)" again yours will be different.) Key in the first Y value (sales in units, second column) 3,000 in this case. (I press a key labeled "Data"; again you may differ).

3 Key in the second X value, 2 in this case, then enter it. Key in the second Y value, 11,000 in this case, then enter it. Continue keying and entering all X and Y pairs until you have entered all five numbers from both data sets.

4 Recall the linear equation coefficient values. (I press "RCL" then "a" and "RCL" then "b", yours will differ) and the values a=0 and b=5000 will be displayed on the screen. You are done; again quite simply.

The result is that the data are best fit by the linear equation $f(x) = 5000x$ or *predicted sales = 5000*number of months*. Plug in x = 9 months and your projected sales at the end of nine months are 45,000 units per month. Won't your stockholders be pleased? Huge sales numbers and it is all mathematically correct. "Mathematically correct" you say? Well, you say it looks that way; after all you did use the well known mathematically accurate linear regression technique. Who could argue with that? Not so fast, "mathematically correct" is a lot like "politically correct" maybe it is right and maybe it isn't. Ask yourself the question: "Does it make sense?" Making sense is too often ignored. The calculations we went through can be done by millions of people. They teach linear regression to hundreds of thousands of students each year. From the looks of things most people believe that linear regression is always enough. But does it make sense? What if we plug in months = 120 and try to project the sales 10 years in the future. The equation works: projected monthly sales are estimated to be 600,000 units per month. But you know that is ridiculous. Your product line is changed every year, so is that of your competitors. Nothing sells for more than a year or two much less ten years. Nobody in the industry ever sells 600,000 units of anything. The model is completely wrong that far out. That is the essence of your problem and the key trick of those who use statistical regression to lie with. Models have a range of values over which they are accurate. Get outside that range and your model can be come increasingly inaccurate up to the point of being ridiculous. You can project anything you want—being "mathematically correct" every step of the way.

So what can you do to predict more accurately and honestly? One thing you can do is to use a more complex mathematical model. Linear regression means a simple linear equation as a model. Most business processes do not follow a linear trend for very long. The next step up is a quadratic model, which means a model of the form $f(x) = a + bx + cx^2$. (Caution: not all calculators that do two variable statistical calculations do quadratic regression. So if you want this capability check before you purchase a calculator.) The squared term allows you to model a bending line and curves that increase faster than a linear curve. Calculating a quadratic regression is as easy as linear regression. For the previous example the only difference between calculating a linear fit or calculating a quadratic fit is to change from keying in a "1" for a linear fit to a "2" for a quadratic fit in step 1 and recall three coefficients: a, b and c. Everything else is exactly the same (again your calculator may differ). For our data set the quadratic regression fit gives this result: $f(x) = -7000 + 11{,}000x - 1000x^2$. This looks considerably different from the linear model $f(x) = 5000x$

and the quadratic model does fit the data much better. Now for the all important question: what do you tell the stockholders? When you substitute the value x = 9 months into the quadratic model, you get 11,000 units as projected sales. Whoops! This is very different from the estimate of 45,000 units for the linear model. This means that sales have already peaked at 23,000 units and will fall off. Not such good news. Still, it makes better business sense. Your companies' new products are known to peak around 6 months. They frequently introduce new models in your industry. This rise and then fall inside of 9 months is standard for a new product. Therefore, the quadratic model fits better and makes better business sense. It is what you should quote to the stockholders.

This is the essential ingredient of modeling. You must know your business area well enough that you can determine how well the model relates to the business process. Also, you must understand how far outside the data region your model is valid. The linear model was a reasonable approximation to the data. Still, it quickly became invalid outside the data region (5 months). However, the quadratic model peaks around 5 or 6 months and then decreases therefore is a valid model for predicting in the 8 to 9 month region. (Aside: this quadratic model predicts negative sales for month 11 and after which means it too is invalid after a point in time.)

Fitting higher order polynomials to data is one way of lowering the residuals and creating a better fit. Look at the sales data for a product line that the dashed line in the next three diagrams denotes. The first diagram has a linear model. A straight line that captures the overall increasing nature of the sales data but is considerably off at many points. The second diagram is a quadratic model. A curved quadratic equation that captures the overall increasing nature of the data. It still matches well at the peak data, where sales peak and then start falling. The third diagram is a complex model incorporating a general equation of the form $a + bx + cx^2 + d\sin(x)$. Sales often peak and ebb during a one or two month period. This is because companies that purchase these items often have set times in a calendar month when they actually order or purchase the goods. Often these peaks and valleys are at the first and fifteenth of a month. The sine function is a model of this business purchasing cycle. A comparison of the three graphs shows the advantage of more complex modeling functions. The first graph fits only an overall upward trend. The second graph picks up additional features like the fall off of sales. The third graph captures all these features plus the business purchasing cycle.

The term "linear regression" sometimes means all three of these models. The term linear actually refers to the model being linear in the unknown coefficients. So any polynomial is a linear model in terms of its coefficients. That third equation with the sine function included is also linear. Just beware of this definition. Many people think of linear as just a linear equation and nothing more. A calculator and standard spreadsheet can probably easily do both linear and quadratic equations. However, that function with the sine term included will require you to use a more sophisticated method.

Sales data is represented by the solid curved graph.

Linear model $y = ax + b$ is the dotted straight line. Notice that it fits only the general upward trend. To fit the downward turn in sales you need ...

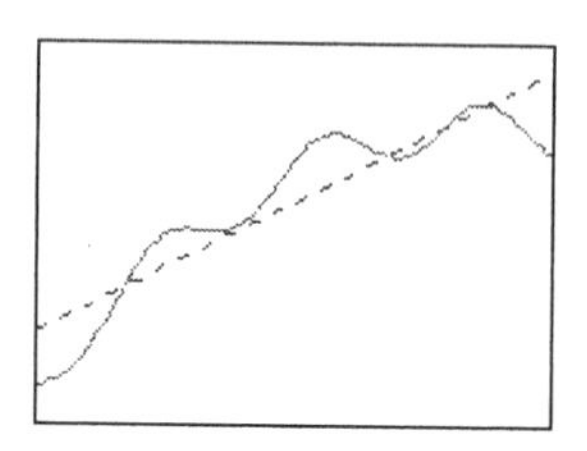

A quadratic model $y = ax^2 + bx + c$ shown by the dotted curve in the second graph. To fit the business purchasing cycle you need ...

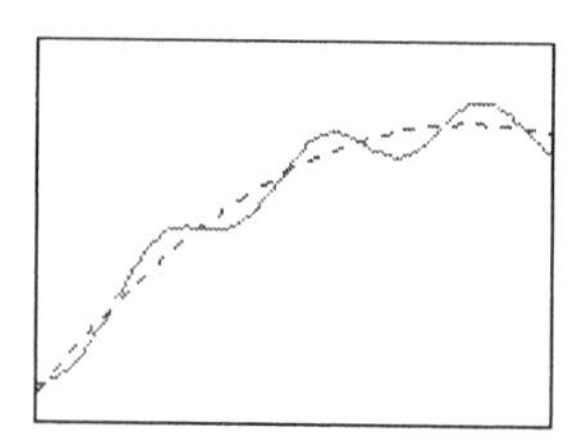

a more complex model given by $y = ax^2 + bx + c + d\sin(x)$ which fits almost perfectly

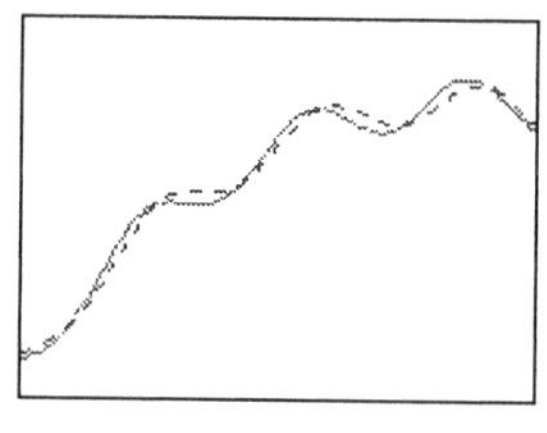

Let's consider another example of regression analysis. This time we use data that depends on more than one variable. The sales-data example, model and graphs were all functions of only one independent variable, time. More than one independent variable leads to a different problem, but multiple linear regression can still solve it if the independent variables are linear terms. Usually you need a multivariate curve fitting program found in statistical software programs to be able to do this. This is what our middle school counselor, Cynthia, had to do. She found that the two most important variables in determining 6th grade academic level math grades were: 5th grade math grades and a standardized test score—two variables. Now, admittedly several more variables could be used. However, using these two variables is somewhat standard and defensible by many studies. Cynthia's problem is the exact form of the equation. She decides upon a linear or weighted sum. Three terms are included, the 5th grade percentage math grade, the deviation from average on the standardized test (this value is *test-47* for her case) and an extra additive offset term. Her model with undetermined coefficients a, b, and c looks like: *grade_6 = a*grade_5 + b *(test - 47) + c*. She uses a statistical program with multiple linear regression capability to fit the historical grade data to this model. She determines the values of the a, b, and c coefficients to be: a = 1., b = .5 and c = -3 which gives a final equation of : *grade_6 = grade_5 - 3 + .5(test - 47).* Cynthia needed to know more about statistics than was needed for that simple business trend example. First, she needed to use her knowledge of previous studies to setup a good usable, defensible general model. Second, she had to pick out the independent variable set. This probably required some correlation analysis beforehand. Third, she had to know how to run a more complicated statistical method than simply running the data through a calculator.

Practice With Correlation And Regression

Some readers might want to practice a little with correlation and regression. The small table below is an excerpt from a larger table in the chapter *Lost in a Large Corporation*. The first column is salary for a set of employees. The next column represents their yearly evaluation (scale 1 low to 5 high) and the last column represents how long they have worked at the company.

Practice with Correlation and Regression

Salary	Yearly Evaluation	Duration
25,000	3	3
47,000	4	10
32,000	2	10
47,000	4	6
54,000	4	2

The correlation values are as follows: salary:evaluation = .79, salary:duration = -.09, and duration:evaluation = -.33. The linear regression between salary and evaluation with evaluation as the independent variable is: *predicted salary = 4875 + 10625*evaluation*. The *Lost in a Large Corporation* chapter goes into detail about how you can use this type of analysis to identify trends in your job in a large business.

Bell Curves

One of the most famous statistical concepts is the Bell curve. We were all subjected to it during our earliest school days, for some reason teachers insisted on grading according to some abstract standard called the "curve." Later this continued and was refined. We seemed to be sorted into the majority on the middle part of the curve (comforting, but hardly gratifying) and those few at the high end of the curve (gratifying, but subject to strange nicknames like "curve wrecker") and those at the bottom of the curve who were treated with a strange kind of pity as if they were out of favor with the gods. Low at both ends and peaking in the middle the bell shaped (therefore its second name) normal curve is one of the most recognized mathematical graphs of all. Even later in school we discovered that not only were our grades scaled according to this curve, but also personal traits. Height, weight, and aptitudes were shaped according to the bell curve. It seemed like a universal plot. To many it still does.

The more knowledge we gained; the more we learned about the world; the more often we saw the normal curve. It occurs in all the social, biological, and physical sciences. Thousands of random variables sort themselves out into the familiar bell curve. There are many examples like height, weight, test scores, economic factors, faulty shipments from an assembly line, the number of traffic accidents on a given day, biological observations, and physical phenomena. Why? Is it really a plot? It is not a plot but it seems like a mystery. The secret is that these quantities are actually a sum of many other smaller factors; a mathematical result called the *Central Limit theorem* applies and explains their tendency to a normal distribution. The central limit theorem states that the *average or sum of a large set of independent measurements* of a quantity is *best* described by a bell-shaped curve. They often call this a normal or Gaussian curve.

What is a Bell Curve?

How can that be? How can one measurement like a person's height or weight or math score on the SAT or the number of heads in a series of coin tosses be the sum of many factors? Consider your math score on the SAT: there are many different questions on the math section, some word problems, some number problems, some geometry, some logic, many algebra and other types. Each has a value (maybe 10 points) and your score is the sum total of all the correct answers. For a second, less obvious example consider: height and weight: a person's DNA with all the different genes and variations within genes largely (not completely) determines the person's adult height and weight. There are several genes that effect height. In a sense each of them can have characteristics that could produce a tall person. If you have five genes that tend to produce tall people and another person only has only two, then on the average you might tend to be taller. (Of course genes' effects are not all equal, but on the average this would be true.) In a sense, your height (and weight) are the effects of sums of genes. That is simple enough; but notice the similarity to the problems from the SAT example. A person's height is the sum of the effects of their genes; a person's SAT math score is the sum of the problems that they get right. Individual coin tosses get summed together; individual gene effects get summed together and individual math problem scores get summed together. The total of coin tosses, the total effect of genes (height) and the total of individual math problem scores (the SAT math score) are all distributed like the normal curve.

So let's concentrate on the simplest example, the number of heads in a series of coin tosses. Note that this is a good example of the Central Limit theorem. This is because the number of heads in m tosses is precisely the sum of the number of heads in a series of m different single-coin tosses. The number of heads is a sum of a set of measurements—measurements in this case being observations about whether the coin shows heads or not. The distribution of heads in a single coin toss is decidedly non normal in appearance. (Zero heads appears half the time and one head appears half the time.) Yet, the distribution of the sum is very close to normal.

Consider this situation: if two coins are flipped simultaneously there are three possibilities: a) no heads, b) one head and one tail and c) two heads. Case b) is twice as likely as either of the other two cases. This is because there are two ways that one head and one tail can arise. These ways are the first coin is heads, second coin is tails and vice versa. The situation is not as clear for say n=6 coins; for example how much more likely is an outcome of 3 heads compared to an outcome of 2 heads? For multiple coins there is a short method of calculating these probabilities that dates back two centuries to Pascal. It involves a table called *Pascal's Triangle* and it leads to a mathematical function called the *Binomial distribution.* The construction of the table is quite straightforward. Each number in a row is the sum of the two (or sometimes one) numbers in the row above it to the right and left. For example the fourth row down (1 4 6 4 1): the 4s are the sum of the numbers 1 and 3 on the row above and immediately to the right and left. The 6 is the sum of the two 3s.

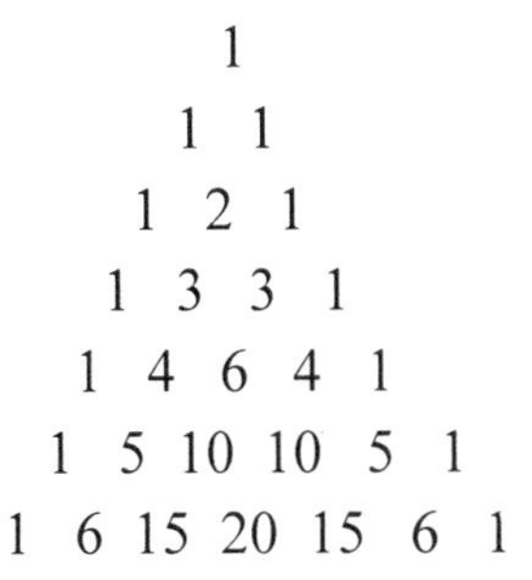
1
1 1
1 2 1
1 3 3 1
1 4 6 4 1
1 5 10 10 5 1
1 6 15 20 15 6 1

Notice the third row: 1-2-1, the numbers represent the number of possibilities for throwing respectively 0 heads, 1 head and 2 heads with two coins. Similarly the next 1-3-3-1 row represents the possibilities for throwing 0 heads, 1 head, 2 heads and 3 heads with a toss of three coins. This says that there are 3 possibilities for throwing 1 head in three tosses—is this true? Yes, consider the act of throwing three coins as two separate acts: throwing two coins, then throwing the last coin. Two situations from the two-coin case can lead to 1 head with three coins: first, no heads in two tosses (represented by the left-hand 1) and the last coin is a head or 1 head in two tosses (represented by the middle 2) and no head on the last toss. The total is 1 + 2 = 3 possibilities. Therefore, the number of possible cases of 1 head in 3 tosses (the 1-3-3-1 row) is 1+2 from the row directly above. This logic can be extended to all other cases. It shows that the numbers in the Pascal triangle represent the number of possibilities. The answer to our previous question about the relative probability of 2 or 3 heads in a six-coin toss can be found by the bottom row (1 6 15 20 15 6 1). There are 15 2-head possibilities and 20 3-head possibilities.

So where is the normal curve in all of this? Consider what each row of the Pascal triangle would look like if we graphed it. The highest(s) number(s) are always in the center, 1s (the lowest numbers) are at the far right and left. The numbers increase steadily from left to center and then decrease from the center to the right. The right and left-hand sides are symmetric. Each row will look more like the normal curve when graphed. Our familiar normal/bell curve is the limit of all the lines in the triangle.

They call the exact statistical distribution formed by the lines within the Pascal's triangle the binomial distribution. It is discrete, not continuous like the normal case. The binomial distribution is not exactly normal for any value of m. Yet, it approaches a normal distribution as m grows. For m equal to 100 or more, the curve is very close to a normal distribution. Our example has involved a special case of the binomial distribution where the probability of the two possible events (heads or tails) is exactly 1/2. The general case is where an event will occur with a probability of p or it will not occur with a probability of 1-p. Pascal's triangle still can be used but the quotient term must change to reflect powers of p and 1-p. We don't need to go into detail here; the resulting curves still approach a normal curve as m grows. The only difference is that the curve is not always symmetric. It can tail off to the right or left depending on the value of p. This is actually the more common situation in the real world. Very few phenomena actually have exactly a 1/2 probability, like a coin toss. Most are like genes or math problems. A specific gene that effects height might appear in only 15% of the population. An ordinary SAT math problem might be solved by 70% of the students on the average. Non-symmetric graphs are more common, but in the limit they look somewhat like the normal curve.

Normal curves represent populations. A single measurement repeated for each member of an entire population. Normal curves are used to compare and contrast populations and groups, not individuals. They represent the variability of a quantity. Mathematically, they are functions of two variables: average (mean) and standard deviation.

How to Analyze Bell Curves

Normal distributions occur everywhere as we have just seen. Also, there are other types of distributions. The average of a population will usually represent the most common occurrence in a normal curve and will then represent the x-axis value at the curve's peak value. A second factor in determining a normal curve's shape is the standard deviation or dispersion of the population. A larger standard deviation gives a curve that is more spread out, a smaller one gives a curve that is more bunched together around the average. These two independent factors, mean and standard deviation, completely determine a normal curve's shape.

So much for theory, let's look at some practical aspects of normal curves. Most of what we say for the Normal distribution holds true for other statistical distributions as well. It's easiest to see with a well-known circumstance in the real world. Peoples' heights form a statistical distribution. More precisely adult female and male heights form two separate nearly normal distributions. The average American female is a little over 5'4" and the average American male a little over 5'9". The standard deviations are also different, slightly larger for the male distribution.

Rely on your own observations and common sense, before doing any statistics. Consider the question of the ratio of the number of men over 5'8" and the number of women over 5'8". No surprise, most men are over that height and only a moderate-sized minority of women are. Now consider the ratio of the number of men over 6'2" and the number of women over 6'2". Big difference here: you will see several men over this height every day but you could go weeks before seeing a woman of this height. The ratio is different. It is not just the relative percentage of each sex that is that height, but the ratio between the two sexes. That is a general pattern: the taller height that you consider the larger the ratio of men-to-women there will be above that height. Maybe one man in fifteen is 6'2" or taller but fewer than one woman in two hundred is, a 15-1 ratio or worse; the ratio for 5'8" is maybe sixty men in a hundred and a fifteen women in a hundred, a 4-1 ratio. A second fact for the pragmatic analyst: if you considered a joint distributions of all adult heights the average would be around 5'7". Also, maybe three fourths of all women would be below that height and maybe three fourths of men would be above. Neither of these facts is surprising. Our common sense observations of adult height distributions support both.

Let's restate these two facts in a general way and consider a general pair of normal distributions on a common graph. (For simplicity we assume that both of the following distributions have the same standard deviation; similar rules are possible for the general case, but they have to be worded more carefully and there are more rules.) First general fact: *If group A has a higher mean than the mean of group B then the proportion of group B that falls above a certain value is small compared to group A and gets smaller as that value increases.* Second general fact: *If group A has a higher mean than group B then group B has a smaller percentage of its members above the mean for the entire combined groups A and B.*

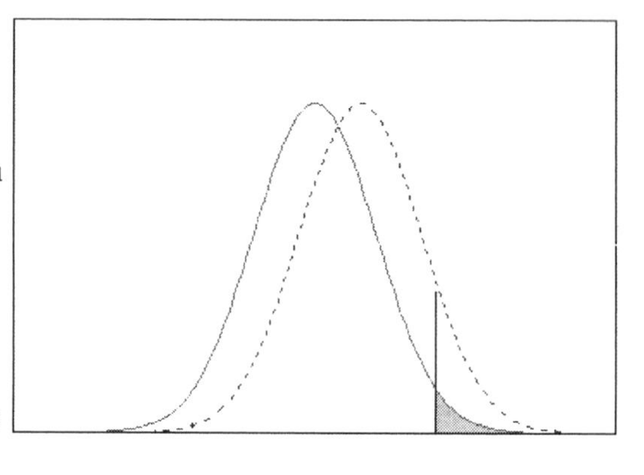

Normal Curves

The dotted curve represents a population that has a slightly higher mean score.

Because of the rapid falloff of the normal distribution, this means that the highest scores in both populations (those to the right of the vertical line) favor the dotted-curve population by a large ratio (4 to 1 in this case).

Notice that these are general statements based upon an examination of two normal distributions one of which has a higher average score. These statements are based on properties of the distributions. Therefore, these two statements hold for any two normal distributions with the same standard deviation. Although we don't show it, they hold for several other common statistical distributions too. They are facts; they are general facts and they apply overall. Also, the standard deviation assumption usually doesn't change these facts. (It is possible to have two distributions where one standard deviation is much larger than the other, but the means are close. Then the fact about ratios needs changing slightly.) Nevertheless, that is uncommon. These two facts are true when two different but comparative normal distributions are compared.

Herein lies trouble—big trouble. Comparative groups with *different average scores* or values are *by far the rule* not the exception. Different sexes, different economic groups, different educational levels, different geographic groups, that differ in seemingly insignificant ways usually differ in other areas too. Normal curves for different groups are usually different and this holds for many different socioeconomic factors. For instance men and women differ in these factors each of which have a related statistical distribution: number of hair follicles, traffic tickets, criminal records, number of words spoken in an average day, life span, amount of phosphorous in the bloodstream, drinking habits, scores on many standardized tests, reflex times, time spent watching TV each night and a thousand more. In socio-medical-economic factors seldom do the two sexes have identical distributions on any one measurement.

And the trouble with this is what? Trouble starts with comparisons between groups; since the distributions usually differ between any two groups—one group must be lower on average. Therefore, the above two general facts will apply: group A will have (perhaps many) more members at the top end of the scores or measurements and more members above the overall average than group B. Trouble is this often looks like class discrimination. Sometimes it is. Sometimes it is not. Reporters, TV commentators and politicians often point out these differences and cry "discrimination." Even worse, often they make the case that group A is holding back group B. Class warfare—it sells—and normal curves have become the weapon. Much worse, sometimes people make the case that group A is superior to group B based on this.

Don't be turned off by a normal curve that you see; after all it is just a mathematical graph. But do, be careful in the conclusions that you draw. Usually, a normal curve represents something measurable that is the effect of many unseen causes. Often analyzing a normal curve is a prelude to a more in-depth analysis of the data using correlation or regression analysis. Normal curves just present data; determining cause-effect or prediction needs more statistical analysis.

Use And Abuse—Statistics For Wannabee Politicians

We have seen how to use correlation, regression and normal curves. Though the underlying theory is mathematical and harder, the actual application of these three techniques is quite straightforward. However, these techniques don't come with a set of directions about how to apply them in every situation. Theory goes just so far; actual statistical applications depend heavily on your application area knowledge. If you have too little knowledge about your area—you can misuse statistics. If you know a lot about your area and want to misuse statistics to prove a point, then you are abusing statistics. This section is about abuse.

Correlation analysis gets abused frequently by noting a strongly correlated pair of factors A and B one of which is undesirable. This is followed by creating an accompanying argument that one factor A is thus a cause of B. Finally, it is argued that to eliminate the undesirable effect B you must eliminate A. This is often used to defend controversial positions. It works like this in practice: a writer does a correlation analysis between the size of city police forces and the violent crime rate in the city. (These two variables are indeed strongly correlated.) That violent crime is undesirable is an understatement. The writer's logic: police presence causes stress, resentment, racism, and fear, all these increase violent tendencies in the cities' population. The writer's conclusion: radically downsize the police forces and cut down on violent crime. Really? This certainly takes care of a major political issue in short order. Yet it does sound a little pat, a little too easy and actually illogical. Most people would see violent crime as the cause for increasing police force size, not the other way around. So they would see downsizing police forces as allowing the violent crime rate to increase. The two factors are strongly correlated. However, a cause-effect relationship is not clear at all. So eliminating one factor (downsizing or eliminating police forces) may do nothing at all to the other factor.

Politicians have been abusing statistics for years; Henry Clay (who has been dead for a long, long time) once remarked: "Statistics are no substitute for judgment." Pity his words are not heeded. Regression analysis is abused frequently by developing models with different underlying assumptions. Modeling the economy is a favorite pastime of politicians and their advisors. Suppose it is an election year; Democratic and Republican candidates are fighting it out for the US Presidency. The economy is the number one issue in the campaign. The incumbent proclaims that the economy has gained during his administration. The challenger points to a recent downturn as proof that the incumbent's economic policy has failed catastrophically. He predicts that next year unemployment will skyrocket, the stock market will crash, there will be civil unrest, rioting in the streets and moral collapse. You get the picture—we have been here before. The next graph shows the economy up to the present and both the incumbent's and the challenger's projections for the future.

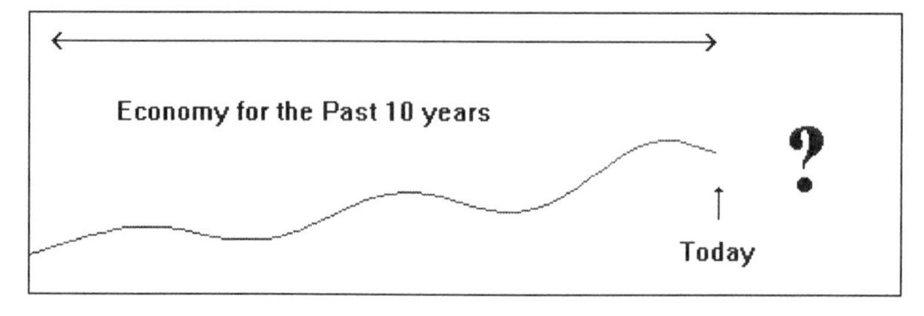

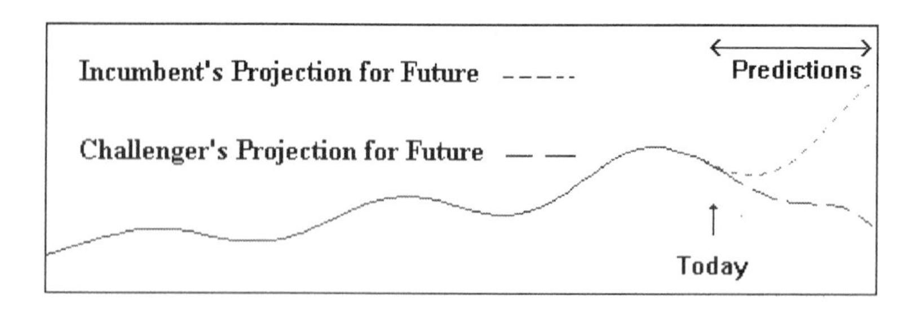

Not much similarity is there? Same data, same mathematics, so what happened? How could you get such different predictions? One way is to limit the scope of the data that you model. The incumbent uses a large time frame (much of it happening before his administration) and dismisses the recent downturn as cyclical. There is a lot of historical evidence for his approach. The challenger uses only the data for the past 2 years since that is when Congress passed the incumbent's big economic plan. The economy peaked after that and started the last downturn. The challenger feels that the incumbent's policy was such a watershed event that the underlying economic forces have changed and changed for the worse. He feels that incorporating earlier data into his model is misleading and dishonest. Who is right? No one knows for sure. The only certainty is that one of these two will win the election.

Normal curves and related statistical frequency distributions are so routinely misquoted, misunderstood and abused that most people are leery of them anyway. That is understandable. A correlation analysis leads to simple statements like: factors A and B are strongly correlated. Regression analysis leads to predictions like: factor A will be at level X in the year 2010. Normal curves represent many facts all graphed together; they are aggregate facts and are more difficult to understand for that reason. Difficult-to-understand can lead to abuse.

A major source of problems with comparing two normal curves is the old apples-and-oranges argument. We often distinguish two separate groups of people. Then we give them a test or make some socioeconomic measurement on them. Later we notice that one group scores better than the other. Often the test scores are normally distributed. So then we have the situation where the group with the higher average has almost all of the top echelon test scores. This leaves the lower group with few or none. This happens all the time. We have high school males compared to high school females on the SAT Math test, white police officers compared to minority police officers on police promotions exams, different sexes and races being compared on the Bar exams or Medical Board exams. Males getting five times more SAT Math scores above 750. Minority officers in some citywide exams not getting any of the top 10% of the police exam scores, therefore not making the cutoff point for promotion. Minority law students not passing the bar exam at the same rate. These are all headlines we have seen before. Often the editorial recommendation is to abolish the test but that doesn't change the underlying socioeconomic factors that are the real cause. And that is the real abuse of normal curve data: focusing on the measurement process instead of the underlying causes.

Checklist

The following checklist is designed to walk you through a statistical regression modeling effort from start to finish.

1. **Define Your Problem** Identify your goal/target/dependant variable(s) data sets; these are the ones that you wish to model or predict. These will be the dependant variables of the model. Next identify your potential input/independent variables. Use your business area knowledge to get as complete a set as possible. This means that your independent variables can determine your dependant variables (e.g., simple math example: if your dependant variable z is such that $z = xy^2$ then you need *both* x data and y data sets to determine the model, just x or y alone won't do).

2. **Gather Your Information** Gather your data together. Choose as uncorrelated a set of input parameters as possible. This means if two of your potential input parameter data sets are correlated strongly to each other, you should strive to use only one of them. (If possible, use your business judgement, this takes second priority to having a complete set of inputs from step 1. The reason for this relates to deeper technical issues involving calculus and statistics.) Split your data into two parts (if practical, sometimes you don't have enough data to do this). The larger part will be the data that is fit to the model to determine the unknown coefficients. The smaller part will be kept aside until the testing phase after the model is done (step 7).

3. **Problem Type** There are three possibilities depending on the number of input and output data sets. Different methods are needed for different problems. If you have one input parameter and one output parameter then go to step 4. If you have multiple inputs and one output then go to step 5. If you have more than one output parameter then go to step 6.

4. **Linear Regression** (one input, one output) This is the classic situation for linear regression. You can fit your output/dependant variable to a polynomial function of your input/independent variable (e.g., the sales trend problem). Either dedicated statistical software or a spreadsheet can be used to determine the model. For low order, linear or quadratic, fits you could also use a calculator. After you are done then go to step 7.

5. **Multiple Linear Regression** (multiple inputs, one output) As a first attempt, form a weighted sum or (multiple) linear regression model—a general model. Here $x_1, x_2 \ldots x_n$ = independent variables and $a_1, a_2 \ldots a_n$ = unknown coefficients. The general model will be: $model(a_1, a_2 \ldots a_n, x_1, x_2 \ldots x_n) = a_1x_1 + a_2x_2 + \ldots + a_nx_n$. Here the predicted value is a function of both the unknown coefficients and the independent variables. Your independent variables will be the cause-data sets. (Some examples are day of the week, temperature, rain and shopping season for donut sales, 5th grade math grades and test scores for the counseling example.) For spreadsheet implementation: see the chapter on Spreadsheets. Go to step 7. Evaluate the fit: if it is good enough then you could stop; if it is not good enough then you should try more complicated models. Study your problem, the input parameters and the fit. Try a more advanced weighted sum model where some inputs or independent variables are composite functions of one or more of your original set of independent variables. Perhaps form x_i^2 a quadratic, for some of the variables. Or perhaps try a product of two different input variables $x_i x_k$ (recall the franchise model had one independent variable, population, multiplied by a quadratic function of another variable, income). Rerun the multiple linear regression fit with this new model. If the fit is good enough, then stop. If not go back and repeat. Study your parameters and your current fit. Try to increase the goodness of fit by increasing the complexity of your model by adding in more terms. This procedure of first setting up a model, identifying the unknown coefficients and then setting up a more complicated model forms a loop that you might repeat many times. If nothing works then stop. At this point, consider using a neural network or a forecaster program.

6. **Multiple Outputs** Use a neural network instead of some complex statistical routine. The statistical alternatives to a neural network in this situation are far more complex than someone should try without a solid background in advanced calculus, advanced statistics, and numerical analysis. Use the checklist in the Neural Network's chapter.

7. **Test Your Model** Run your fitted model with the testing data that you set aside in the first step. If you are satisfied then you are done. If you aren't satisfied then you can try a more complicated linear model by going back to step 4 for linear regression or step 5 for multiple linear regression. If nothing ever works well enough then you might have to give up on regression at some point. You have two choices: nonlinear regression, which almost always requires calculus and statistics or you

can use a neural network which can do the same things as both multiple linear regression and nonlinear regression but is slower to run on a computer. If this is how your problem turns out read the Neural Network chapter.

8. **Predict with Your Model** Now that you have completed a model that you have tested successfully against your data and now that you are satisfied that it fits the data well enough for your needs. Now you are ready to predict with it. Normally, this means that you have some values of the independent variables that represent a scenario that has not occurred yet and you want to know how it will turn out. Simply plug in the values of these independent variables/inputs and evaluate the model. This will give you a prediction and only a prediction. If you want a good set of error bounds or a confidence interval around the answer you must delve deeper into the study of statistics or simulations.

Chapter 6

The Essential Spreadsheet

It Makes The Business World Go Round

Meetings featuring presentations with pie charts, bar graphs and tables of processed data are common in the American workplace. American industry can't plan without its charts and viewgraphs. Different scenarios, predictions, business models, risk assessment and industry trends are all generated by computers then presented to groups of managers and workers. Sound familiar? The mainstay program that generates most of this is a spreadsheet. A spreadsheet program is the most widely used data analysis software tool in business today. They are versatile: they calculate, they sort, they search, they present data, they analysis alternatives (what-if studies) and they can develop models too. They easily link to word processors, graphics programs, database programs and the specialized prediction software that we discuss in this book: regression and neural networks. You have to know how to use spreadsheets to be successful in business today (or be able to hire a close assistant who knows spreadsheets).

This chapter is *not* a general introduction to spreadsheets. You can learn that elsewhere. There are many excellent introductory books, videos and classes about spreadsheets. This chapter is about how to use spreadsheets to do two modeling tasks in our fundamental theme of *data->model->plan*: move from data to a model: data->model or move from a model to a plan: model->plan. When we discuss regression or neural networks as good prediction tools we aren't excluding also using spreadsheets along with them. Spreadsheets can be used after a model is developed by regression or neural networks. In this *model->plan* case spreadsheets can take the previously developed model and perform what-if studies or optimal policy calculations with these models to develop the most profitable plan. Also, spreadsheets can be used directly with data to develop a model without separate statistical software or neural networks in the *data->model* case. This chapter is in four sections. First, is a general overview of spreadsheet capabilities. Second, is a simple *data->model* example revisiting the bakery example. Third, is an advanced *data->model* example involving a couple that invests in a franchise. They need to decide the best location for the store. Fourth and last, is a *model->plan* example, a what-if study involving the optimal amount of over baking that should be done above the model's prediction. This safety margin is necessary to maximize profit and keep disgruntled customers at a minimum.

Spreadsheets—What Can They Do For You?

First, we look at the basic groups of functions and capabilities that modern spreadsheets offer. *Text functions*—spreadsheets can do searches for text strings and replace designated text strings with other text strings. They can also sort or index on different variables. They are very similar to word processors and databases in this respect. *Logical functions*—basic boolean operators like AND, IF and NOT are provided. These are very important since they allow you to create spreadsheets functions and other simple branching operations. These allow you to put together more complex operations. *Statistical functions*—linear regression, standard deviation, several statistical distribution functions and many more are available for processing your data. These are very useful if you need to do standard statistical calculations and don't want to buy and learn a separate standalone statistics package. *Math and trig functions*—many standard functions are provided: powers, sine, cosine, exponent and others. *Presentation capabilities*—charting functions, some drawing functions and others are standard. *Programming language capabilities*—programming languages like Basic, C, C++, or Java are widely used because they can pull together a variety of operations in a single program. Modern spreadsheets (like the new releases of MS Excel) allow you to code macros in a programming language (Visual Basic for Excel). Then you can run these within the operation of the spreadsheet. If you can program this opens up new vistas for you. You can customize your spreadsheet operations and add new functional capabilities. Note at this point you are getting past our rating of spreadsheet difficulty at the 3 level. Programming is usually a more difficult skill to master. Still having this capability included within a spreadsheet is easier to work with then writing a program outside the spreadsheet and then linking it to the spreadsheet operation.

Data->Model With A Spreadsheet (Simple Case)

Let's revisit the donut bakery example. Our donut shop owner Maria had a year's worth of records from her first year of business. Recall that Maria was very conservative and over baked everyday to avoid alienating customers by running out of freshly baked goods. The second year she realizes that she is wasting too much money that would be profit going to her needs like house additions or European vacations with her mother. Her records are a computer file with 360 rows of data, 5 columns to a row. From this she needs to develop a predictive model that will accurately predict tomorrow's demand for donuts. In previous chapters, we saw several different models and baking strategies based on just business sense, multiple linear regression and neural networks. In this section we consider several different models based upon different levels of spreadsheet analysis. These won't necessarily do better than the ones we saw earlier. Nevertheless, they do illustrate how simple spreadsheet analysis can lead to useful business plans. Let's look at several models that predict the next day's demand for donuts.

Model #1: use only the day of the week as a relevant variable. For example, average all Monday sales together and use that average as the model or predicted value for next Monday. (In Excel you might use the "DAVERAGE" function to do this.) Model #2: use day-of-the-week and get as close to tomorrow's predicted temperature as possible. One way to do this is to average all data from that same day of the week that had a temperature within plus or minus 5 degrees. For example: tomorrow is Monday, the predicted temperature is 40°. Use the spreadsheet with functions like DAVERAGE and AND along with inequalities like"<=" then search for all Mondays with temperatures between 35° and 45°. With these data you can use spreadsheet functions to average the sales for these days. Then you can use that average value as the prediction for next Monday's demand. Model #3: use the same variables as model #2 but fit a linear regression curve to demand as a function of temperature for Monday. (It is similar for other days of the week.) Then substitute in temperature = 40° and this will be a prediction of tomorrow's demand (a function called FORECAST in Excel might be used here). Model #4: use model #3 as a baseline but then separate out (with logical operations) days of different shopping seasons and by comparing demands of the same day of the week with similar temperatures you can get a multiplicative factor that represents the higher volume of shopping done during that season, like Christmas season and Labor Day weekend. Model #5: the same as model #4 but do another factor for rain.

You see the pattern. Each model is an enhancement; it will give more accurate predictions at the expense of more functions and logic within the spreadsheet. The first simple model only requires using the spreadsheet averaging function. The fifth model required: an averaging function, a regression fitting function, two or more specific multiplier cells with related formulas, logic that searched for and separated out subsets of rainy days and different shopping seasons, other logic that averaged these subsets and divided the results then stored them in the multiplier cells—quite a bit of logic and manipulation. For simple data sets it's probably easier then learning a neural network software system; for moderate-to-difficult data models another approach like a neural network might be better. You can make difficult models with spreadsheets although it takes more knowledge of math. We will see one in the next section: a nonlinear regression method done by a spreadsheet.

Data->Model With A Spreadsheet (Advanced Case)

Our general theme has been data->model although we have developed three types of models: sets of rules, math functions, and computer programs. When developing a math function model (like for regression) there is usually a step in the data->model process where you define a general model or class of functions. The model is then fit to the data and the undetermined coefficients of the general model are determined by the fitting process (regression of some form or genetic optimization). For example: you might consider your data, the processes and principles that it represents and decide that the general model for the process is $y = a + bx + cx^2$, a quadratic model with undetermined coefficients a, b, and c. After fitting this model to your data by regression you might get: $y = 2 + 3x + x^2$ as the specific model. The key point: if you can determine the general math form of the model beforehand—then do it. It is a big help to the fitting process if you know this extra information. However, often this is not possible. Our next example shows a case where you can deduce the general model from business principles. Then you can then use advanced features of a spreadsheet to determine the undetermined coefficients of the model.

Some business decisions are so important that you can't retrace your steps after making them. Some business decisions are so obscure that you don't know any really good rules-of-thumb or general knowledge. It is at these times that you must mathematically analyze the available data and make a decision based on that. Here is a case where you can apply statistical analysis or neural networks and both will give good results. Yet rules or business sense don't lead anywhere. Here is a case where just knowing what you can do with the data is most of the battle.

The next example is where the people involved have a once-in-a-lifetime decision that they must make. If they don't want to learn the math to do it themselves, they would be well advised to hire a consultant. Just reading this book and knowing that they can solve this type of problem is very important. It will make Bob and Josie more than a million dollars in a few years. More important, it will spell the difference between the success and failure of a dream of a lifetime.

Bob and Josie have spent most of their married lives working for other people. They have always worked well together on projects around the house and have especially enjoyed the togetherness and common interests generated by the planning and execution of their plans. They want to go into business together. With this goal they read franchise magazines and sent away for literature. Finally they decided to purchase local area franchise rights in a new company that has 20 franchises in 9 states. For their $200,000 franchise fee they get exclusive rights to operate a company store in their designated area. Also, they get statewide publicity, equipment, advice, training, and data from the parent company.

Bob and Josie spent their entire savings on a franchise in their city.
A make-or-break decision is: what is the best location for the store?
How can data analysis help?

Bob: "I am tired of working for others. There are twenty years left before I can retire and then what? My job is OK but its not a dream job, I don't even think of it as a career—just a job. Josie and I have talked for years about having our own business and working it together. This franchise is our chance. Yes, it's expensive and yes, it is risky. We could lose everything; it took all our savings. Still we can make a lot of money if we can just get a good location to start in. We can run the business but we must have a customer base to start with."

Josie: "We work different jobs, different hours, don't have kids, no real hobbies or anything. We just get together some nights for TV or bowling or time with our friends. We have saved a lot of money over the years by driving used cars and not taking vacations or spending foolishly. And we have always had this dream of being in business for ourselves. Without kids or interesting careers that is about all we had. Sometimes that dream has kept us going through all the extra hours, mean bosses and dreary surroundings. At some point we have to try to catch that dream or just kind of dry up and blow away. I love Bob too much to want that."

Bob: "We have read a lot of books by experts, but the advice contradicts each other. We have several places to put our franchise store—but which is better? Do you aim for richer areas at the expense of smaller population density and how do competitor stores figure in the equation? Is the location on 22nd and Elm really $2000 a month better than the one at 12th and MacArthur; it is a newer mall in a richer area, but doesn't have as large a population within a 5-mile radius? How can we tell? Or how about the two others? Once we decide we can't easily switch locations; all of our savings are tied up in the franchise fee and startup costs. The store location will determine our chances of making it. How can we tell which is best?"

Facts: there are more than 2,000 franchise systems in this country. People invest their savings into franchises in hopes of making big salaries and being their own boss. After years of working for bosses with their own managerial peculiarities and personalities many people will risk almost anything to be on their own. Franchises offer this possibility. Nevertheless, franchises are risky and even worse chances of making large incomes are small. It has been estimated (Ann Dugan, business professor at the University of Pittsburgh, quoted in the Washington Post) an average franchisee can expect an annual salary of about $35,000. Many make less or work free in effect. Profit margins are often small. Several chains (some of whom didn't make it, from the Washington Post again) are named like: Taco Eds, Captain Crabs, and Catfish station. Of course McDonalds, Taco Bell, and others are often doing fine for their owners and many people have become millionaires.

Checking around Bob and Josie find four shopping centers or malls in their area that they think will be a good location for their new store. The rents vary greatly; they are larger overall for the more upscale malls. Logic: it seems that to place their store in the center of richer or more densely populated areas would lead to more sales, hence higher profits. Their problem: which of this four areas will generate the most profit for them year after year. The four possibilities are in different income areas and have different population densities. Which of the four is best? They can't redo this decision later because the startup cost is too high. They will be stuck with their decision for a long time. It is probably a make-or-break decision with the hopes of a lifetime riding upon it.

Data: the parent company does provide some data for them. Twenty of the stores have provided a year's gross sales figures to the parent company. In addition the company has provided demographic information in the 5-mile radius surrounding area for each store. Therefore, for each store they know: average income, area population, number of close competitors. They can do a lot with this. The following table is a sample of this data.

Data From Other Franchise Stores

Area Income	Area Population	Competing Stores	Gross Sales
57,000	67,000	1	407,000
75,000	98,000	4	172,000
44,000	56,000	2	120,000
40,000	40,000	3	51,000
66,000	56,000	2	249,000
70,000	75,000	4	180,000

Bob and Josie have checked out the four different malls within their franchise area. Also, they have obtained the area income, population density, number of competing stores and the yearly rent at each of these locations. The company told them that their profit margin is about 30% of gross sales. With this fact they can estimate profit if they can develop a model to predict gross sales at each of these four possible store locations. Their data looks like this:

Possible Locations For The New Store

	Area Income	Area Population	Competing Stores	Yearly Rent	Profit
A	50,000	80,000	1	24,000	?
B	60,000	70,000	1	30,000	?
C	70,000	90,000	3	42,000	?
D	80,000	50,000	2	48,000	?

Look at this table and recall that the decision of a lifetime is riding on your answer. One of these locations makes less than a living wage for Bob and Josie—go broke slowly. One makes a little more than a living wage—hang on by your fingernails. One goes broke quickly and one makes them a six-figure income, the kind of living they having been dreaming about. They get one chance. Take a guess, answer at the end of this section.

We have data, now what do we do? First step—think. Recall the bakery problem. There we had general principles like: 1) less demand at the beginning of the week because dieters often start then, 2) colder weather—more sales, 3) rainy weather—fewer mall customers. At that point we put the data into a neural network and ran it. That approach would work here too. Neural networks can infer relationships like the above principles from just data. However, a better approach might be available for this franchise problem. We can develop a general math model for this situation. (Aside: this is seldom the case in complex business decisions since there are so many parameters that effect the outcome. This is more common in engineering where only a few physical principles effect the outcome. Nevertheless, this is the *best* way to do modeling *if you can.*) Let's think. Let's think in general terms about the effect of income, population and number of competing stores on gross sales. *Population density*—suppose there is a store in an area with 20,000 people within shopping distance. There is a second store in an area with 40,000 people. Both shopping areas (a 5-mile radius) have exactly the same income level and number and kind of competing stores. Question: how would gross sales between these two areas be related? A very good estimate is that twice as many people would mean twice as many sales. This means that population density is probably a linear term—a multiplicative factor in a math function model of sales. *Competing stores*—assuming that there is a constant demand for your product then if you have one equal competitor then your share of the market is 1/2 if you have two competitors it is 1/3 if you have 4 competitors then it is 1/5. The math term that represents this is: $1/(1+s)$ where s is the number of competing stores. This is an approximation since some competitors might have more of an impact than others; but still it is a logical general math relationship and is a good addition to our general formula. (Aside: also later we will put in other unknown variables that can compensate to an extent for any differences between this general term and reality. This is standard practice when you don't have an exact equation.) *Income*—This is the tough one. What is the dependence of sales as a function of the average customer's income? Will a customer with a $100,000 annual income buy twice as much of your product as one who makes $50,000? Or is your product too "middle class" and the richer customers won't be interested, they will tend toward more expensive products that are not direct competitors with your products? You aren't sure. The parent company doesn't know either, however, they believe that their product is more attractive to buyers with incomes between $40,000 and $80,000. Your product is not a necessity so sales also fall off toward the lower end of the income scale. OK, where does all this leave us? If we knew the exact demand for the product as a function of income, the graph would start low for lower incomes. Then it would peak in the income region somewhere between $40,000 and $80,000. Last it would then fall off to lower values for high incomes. Essentially, the graph has a peak in it; we don't know the location of this peak and we don't know any values of this function for any income. Does this really help? Yes—this one fact, this one idea is actually a tremendous help. It allows us to assign a general math factor to the income effect. We don't know the exact form but we are willing to settle for a good approximation to the shape of the function. Several general math functions start low, peak then fall off: trig functions like sine and cosine, polynomials, and various Fourier series. We will take the simplest form: a quadratic polynomial function that has the general form: $f(x) = a + bx + cx^2$ where x = income and f(x) = demand for your product. This quadratic function will rise, peak and then fall; review the sales trend example in the Statistics chapter.

The final general model includes all three of these terms as multiplicative factors. Multipliers because the income function estimates how much one potential customer might want. Then you must multiply by the number of potential customers (the population in a 5-mile radius). Finally, you multiply by a factor that takes into account the competing stores, that 1/(1+s) term. We also include an unknown coefficient multiplier that scales the level of the function to the actual sales. Our final general model: *model(k,a,b,c) = k * population*sales due to income level*market share because of competition* $= k * y * (a + bx + cx^2)*1/(1+z)$ where x = income level, y = population and z = number of competing stores. There it is a general math model. For a concrete example we will consider the case of MS Excel. Excel has its own equation solving method called "Solver" that will be used to develop a specific model from the above general math model.

A spreadsheet modeling approach using Solver could work like this. First, you would input the columns of data from the parent company into the spreadsheet. Second, you could type in Basic modules representing your model function and the least squares function. Third, you can assign an initial estimate of your model's unknown coefficients to changing cells in the spreadsheet (here k, a, b, and c). Fourth, you could evaluate the difference between the model and the data for each line of data. Then you designate a cell to hold the summation of all these differences squared. This summation would be the least squares function that needs to be minimized. Fifth, you can run the Solver program with that cell in the fourth step as the goal/target cell and the unknown coefficients of the third step as the changing cells.

That is the plan; now let's go through these steps in more detail. A sample spreadsheet is listed to start with. The following is a small Excel spreadsheet that will run this data and this model. The functions are not visible because the spreadsheet has evaluated them.

		Data from Parent Company					
	Income	Population	Competing Stores	Gross Sales		model(k,a,b,c,x,y,z)	
	57000	67000	1	407000		385000	
	75000	98000	4	172000		209000	
	44000	56000	2	120000		113000	
	40000	40000	3	51000		34000	
	66000	56000	2	249000		229000	
	70000	75000	4	180000		178000	
		Values of coefficients					
	k	a	b	c			
	1.6	-31	0.001	-9.3E-09			

Next we will fit this model to the data table supplied by the parent company. "Fit" is used in the sense that we saw before in the chapter on data analysis: where the value of the model function is subtracted from each data point, this difference is squared then all the squared differences are summed. Then the summation is minimized. The summation is a function of the unknown coefficients of the model function (k, a, b, and c). We will let the computer search for a set of values for k, a, b, and c. These values minimize the sum of the squares of the differences between the model math function and the data which are gross sales here. In symbols: minimize $k\ [data - model(k,a,b,c)]^2 = k\ [data - (k * y *(a + bx + cx^2)*1/(1+z)]^2$. There are various ways that the computer can do this.

You need to implement these two functions, the model function and the summation function, in the spreadsheet. Notice that the data column is labeled "Gross Sales" in the spreadsheet and the model column is labeled "model(k,a,b,c,x,y,z)." The function "SE" (squared errors) subtracts these columns from each other then squares the results row by row. Finally the spreadsheet summation function sums all the squared differences. For this case you could implement in a Visual Basic module and access these functions from the spreadsheet. The Basic code for the model function and the SE function are shown below.

```
Function Model(k,a,b,c,x,y,z)

        Model = k*y*(a+b*x+c*x*x)*1/(1+z)

    End Function

Function SE(data,k,a,b,c,x,y,z)

        temp = data - Model(k,a,b,c,x,y,z)

        SE = temp * temp
    End Function
```

Solver is a general linear/nonlinear equation solver/optimization program included as part of a full installation of Excel 5. It starts by asking you what is your target or goal. Keeping with normal spreadsheet usage, functions and variables are stored in cells. So the program will ask you what your target cell is (dependant variable). It will also ask whether you want to solve for a max or min or value of the function. Finally, it will ask you what cells can change (independent variables). The summation function calls the model function; the goal/target is to minimize the summation function; the changing cells are k, a, b, and c and for this example there aren't any constraints. When you run Solver it plugs in values of your changing cells and then recalculates the spreadsheet. This in turn recalculates the value of that summation. Then it tests whether the summation value is less, therefore better, than it was for a previous set of values for the changing cells. This is a standard method. Each trial run with a new set of independent variable values (changing cells) is called an iteration. The results of iterations are saved and compared. Calculus methods, included in Solver, give directions for choosing values so that the succeeding iterations are improvements. Eventually Solver stops; it may have an answer, it may not. Its "answer" might be what you want, it might not be. You aren't necessarily done yet. Rerunning Solver with a different set of values for the changing cells might be necessary. This might give you a better solution. This is often done several times. There are deep technical reasons why you can't always start a program like Solver running, then come back in five minutes for an answer. These reasons get quite involved and will be touched upon in the Calculus chapter. For this example Solver did converge to a specific model.

How did it turn out? Which option was best? The best answer is B; Bob and Josie make a six-figure income and their dream comes true. If you answered A: then they hang on, meet all expenses and have some left over. Answer C means going broke slowly and if you answered D, well then it was a quick trip to bankruptcy court. Not obvious was it.

Sometimes you absolutely have to analyze data and form a prediction model. If you can't do it yourself then you must hire a consultant. Even after reading this example, you might not feel that you want to do it for your own business problem. At least you know that you need help. Just knowing that data analysis can do this is crucial. That is the real point of the example.

Model->Plan With A Spreadsheet

Spreadsheets used with preexisting models can perform What-If analyses. This type of study is fundamental to optimal planning with models. A What-If study considers what would happen if the variables had a certain set of values. They refer to a set of variable values as a scenario. Models reflect reality; therefore, by trying different scenarios you can see what would happen if you tried different plans in the real world. If you have a method of comparing plans (profit is usually a good measure) then you can use a model as a trial run to see which plan is best. Spreadsheets are useful here; that Solver program we used in the previous section we can also use here.

Let's revisit the donut shop. We have several models all of which predict tomorrow's demand for donuts. "Predict" is the key; no prediction will be perfect, if our model predicts sales of 1400 donuts we are not surprised when the actual demand is 1200 or 1600. A random component exists that we can't model. However, we have to deal with that random component. Say we bake 1400 and the demand was only 1200. We waste the ingredients and baking expenses for those 200 extra donuts, $30. Say we bake 1400 and the demand was for 1600. We lose 200 donuts worth of sales, $80. Even worse, we annoy many hungry customers who were anticipating munching one of those 200 donuts that never got baked. Over baking costs; but shortages cost far more. Common sense demands a safety margin of over baking above the number predicted by the model. But how much over baking is best?

We use a What-If study to choose an optimal percentage of over baking. Our method is to use the data from last year along with our model. We use the weather, day-of-week and season information for each day and then use the model to predict how many donuts we would need (note: of course we already know how many donuts were needed). Next we assume that we baked a certain percentage above the model's prediction; this gives us a plan based upon a variable, the percentage of over baking. (Aside: this might get confusing; to clarify this, think of one plan as baking the model prediction plus 5% each day, consider a second plan as baking the model prediction plus 10% each day. Now the 10% plan is safer and will result in more donuts actually being sold because there will be fewer shortages over the span of a year. However, the 10% plan will waste more donuts per day than the 5% plan and that is an additional cost.

So which is best? Well the way to find out is to try both plans out with last year's data. A more general way is to have a variable amount of over baking that can vary from 0% to say 20%.) We compute the profit for each plan and we then decide which possible plan is best. (Recall Maria over baked by a big margin all year so her sales data should reflect actual demand; she didn't have many shortages if any.)

This sounds like a good method but how do we compute the best over baking factor? Solver will help. First, we assign a cell that represents the percentage of over baking. Then we construct functions that compute the amount of sales, waste and shortage. These functions are based on using this percentage of over baking with the model's predictions against last year's data. Then use Solver with the over baking factor cell as the changing cell. Use the cell that holds the profit function as the goal/target cell. Also, set up a constraint that requires that the cell that holds the total shortage for a year be less than some reasonable value (like 10,000 donuts for a year). Then run Solver and it can find the best possible over baking factor.

Checklist

Should you use a spreadsheet to develop a model or should you look to statistics, neural networks or something else? That is a tough question but one that you have to face. Read the other examples in this book and consider the other methods. If you decide upon a spreadsheet as your primary modeling tool then the following checklist will help you develop a mathematical model.

The following checklist is designed to walk you through a regression modeling effort with a spreadsheet program. The checklist is similar to the statistics checklist. Several additions are included which are only applicable to spreadsheets. Also included is a discussion of what to do if you already know the model.

1. **Define Your Problem** Do you have or can you construct a general math model of your process? If yes, refer to the franchise example and see if you can use the spreadsheet to identify the unknown coefficients. Follow that example and skip the rest of this checklist until step 7. You will also need to split your data into two parts as discussed in step 2. If no, then continue. Identify your goal/target/dependant variable(s) data sets; these are the ones that you wish to model or predict. These will be the dependant variables of the model. Next identify your potential input/independent variables. Use your business area knowledge to get as complete a set as possible. This means that your independent variables can determine your dependant variables (e.g., simple math example: if your dependant variable z is such that $z = xy^2$ then you

need both x data and y data sets to determine the model, just x or y alone won't do).

2. **Gather Your Information** Gather your data together. Place them in the spreadsheet in a way that the different types of data can easily be distinguished from each other. Next consider the potential input/independent data sets and perhaps run correlation calculations between pairs of them (use the built-in spreadsheet functions for this, it is easier). Choose as uncorrelated a set of input parameters as possible. This means if two of your potential input parameter data sets are correlated strongly to each other, you should strive to use only one of them. (If possible, you must use your business judgement. This takes second priority to having a complete set of inputs from step 1. The reason for this relates to deeper technical issues involving calculus and statistics.) Split your data into two parts. (If practical, sometimes you don't have enough data to do this). The larger part will be the data that is fit to the model to determine the unknown coefficients. The smaller part will be kept aside until the testing phase after the model is done (step 7).

3. **Problem Type** There are three possibilities depending on the number of input and output data sets. Different methods are needed for different problems. If you have one input parameter and one output parameter then go to step 4. If you have multiple inputs and one output then go to step 5. If you have more than one output parameter then go to step 6.

4. **Linear Regression (one input, one output)** This is the classic situation for linear regression. You can fit your output/dependant variable to a polynomial function of your input/independent variable (e.g., the sales trend problem). For a linear equation the spreadsheet will probably have a built-in function that you can use. For higher order polynomials you might have to provide the equations within the spreadsheet. After you are done go to step 7.

5. **Multiple Linear Regression (multiple inputs, one output)** As a first attempt, form a weighted sum or (multiple) linear regression model—a general model. Here $x_1, x_2 \ldots x_n$ = independent variables and $a_1, a_2 \ldots a_n$ = unknown coefficients. The general model will be $\text{model}(a_1, a_2 \ldots a_n, x_1, x_2 \ldots x_n) = a_1x_1 + a_2x_2 + \ldots + a_nx_n$. Here the predicted value depends on both the unknown coefficients and the independent variables. Your independent variables will be the cause-data sets (e.g., day of the week, temperature, rain and shopping season

for donut sales, 5th grade math grades and test scores for the counseling example). For spreadsheet implementation, assign each coefficient a cell with an initial value. Then use a solver-type program to minimize the sum of the squared error between the model and the independent variable values. Go to step 7. Evaluate the fit: if it is good enough then you could stop; if it is not good enough then you should try more complicated models. Study your problem, the input parameters and the fit. Try a more advanced weighted sum model where some inputs or independent variables are composite functions of one or more of your original set of independent variables. Perhaps form x_i^2 a quadratic, for some of the variables. Or perhaps try a product of two different input variables x_i x_k (recall the franchise model had one independent variable, population, multiplied by a quadratic function of another variable, income). Rerun the multiple linear regression fit with this new model. If the fit is good, enough stop. If not then go back and repeat. Study your parameters and your current fit and try to increase the goodness of fit by increasing the complexity of your model by adding in more terms. This procedure of first setting up a model, identifying the unknown coefficients and then setting up a more complicated model forms a loop that you might repeat many times. If nothing works stop and consider using a neural network or a forecaster program.

6. **Multiple Outputs** My advice is to use a neural network instead of some complex statistical routine. This probably means that you can't do your problem entirely within your spreadsheet program. A possible alternative: split your problem up into several separate problems—each a single output problem. Next form a model for each and see if that helps you. It might turn out that the models are close enough to each other for your purposes. (Sometimes this works, but often it doesn't and you have to go to a different method, like a neural network.)

7. **Test Your Model** Run your fitted model with the testing data that you set aside in the first step. If you are satisfied then you are done. If you aren't satisfied then you can try a more complicated linear model by going back to step 4 for linear regression or step 5 for multiple linear regression. If nothing ever works well enough then you might have to give up on regression at some point. You have two choices: nonlinear regression, which almost always requires calculus and statistics or you can use a neural network which can do the same things as both multiple linear regression and nonlinear regression but is slower to run on a computer. If this is how your problem turns out read the Neural Network chapter.

8. **Predict with Your Model** Now that you have completed a model that you have tested successfully against your data and now that you are satisfied that it fits the data well enough for your needs. Now you are ready to predict with it. Normally, this means that you have some values of the independent variables that represent a scenario that has not occurred yet. You want to know how this scenario it will turn out. If you have these values stored in the spreadsheet or have the spreadsheet dynamically linked to a file containing them, then you can let your spreadsheet model evaluate them. This will give you a prediction and only a prediction. If you want a good set of error bounds or a confidence interval around the answer, you should delve deeper into the study of statistics or simulations. Also, your spreadsheet might have a Monte Carlo capability or have an available add-on that could help you do these sensitivity/confidence interval studies.

Chapter 7

Calculus

Do You Need It For Your Data?

Did you skip taking calculus in high school or college or did you take it but come away with a vague feeling that you just didn't get it? Don't feel alone. Did you ever listen to some math whiz pontificate about calculus; how great it is, how enlightening it is and somehow feel that you missed something? Don't feel alone. Did you ever pick up an engineering or physics textbook, turn the pages and simply marvel at all the calculus-type equations and wonder how anyone could understand it? You are not alone.

Calculus has a mystique all of its own. Available at the high school level, they most commonly teach it in college. It is the first of the truly abstract, theoretical math courses. Yet part of its mystique is its applicability. By reputation, it can do wonderful things. It can solve great problems, enlighten and enrich those select few that understand it. A mathematical merit badge, a certificate of excellence, the keys to the mathematical kingdom, it has the reputation of being all these and more. Added to these perceptions is one more. Calculus is tough, very tough. Only the best and brightest can understand it. So if you have passed it then you are elevated in many eyes.

I studied calculus for years and taught it for more years. I did mathematical research and published papers in area that could be called advanced, advanced calculus. (Actually it is called functional analysis. You need a year and a half of calculus, another year of advanced calculus, plus several other advanced math courses all before you can even enroll in it.) During the 70s and 80s, I programmed calculus methods into computers to solve complicated engineering problems.

Today—I don't even use calculus. Yet, I do use software that uses calculus—modern software incorporates a lot of calculus that is invisible to the user. You don't have to understand any calculus to use a lot of that software.

Calculus is by far the most discussed, the most analyzed and the most funded college math course. They reform calculus courses constantly; many college classes are research vehicles for researchers that are trying different teaching methods. No math course ever caused as much contention as calculus. College teachers at conferences debate the merits of different approaches endlessly and often bitterly. But this chapter is not about learning calculus; it is not about the wonders of the infinitesimal. Instead, we want to see what calculus can do to help us analyze large data sets. This chapter is also about software alternatives to calculus.

Derivatives and Integrals

Three main topics in beginning calculus are the infinitesimal, derivatives and integrals. A study of infinitesimals is the theoretical foundation for calculus. It is arguably the most difficult part of calculus. Yet infinitesimals are a means to an end: they are used to define the derivative and integral. This is a how-to-do data analysis manual, not a textbook. So we are going to bypass infinitesimals and move straight to the applications of derivatives and integrals.

Both the derivative and the integral are functions associated with another function. For those that haven't any background in calculus some examples will illustrate. Consider the function $f(x) = x^2$. The function $2x$ is the derivative of $f(x)$ and the function $x^3/3 + c$ (c is a constant) is the integral of $f(x)$. For the function $f(x) = \sin(x)$, the function $\cos(x)$ is its derivative and $-\cos(x) + c$ is its integral. For the function $f(x) = 1/x$ its derivative is $-1/x^2$. Its integral is $\ln(x) + c$. The actual process of deriving the derivative and integral of a function can be very hard and lengthy. This process is the subject of much of the first year of calculus. But we don't need it. There is software out there that can find both derivatives and integrals better than anyone except math professors and assorted math geniuses.

The secret of calculus is that these two derived functions, the derivative and the integral say important things about the original function f(x). Now since math models are functions, it follows that derivatives and integrals of these model functions can tell us important things about our models. In turn, we develop important facts and insights about anything we model mathematically. The list includes the economy, physics, biological structures, transportation systems, chemistry, engineering, health care, astronomy, the stock market and a thousand more.

What important things can a derivative and integral tell us? A derivative gives the value of the slope of a function or its rate of change at each point.

Often models are functions of several independent variables. The slope of the curve with respect to a variable is the same as the sensitivity of the curve with respect to that variable. It tells how much impact that variable has on the model. In addition, two important points on function graphs are the maximum or minimum points. At these points the slope of the curve, hence the derivative, is zero. Therefore, by solving the derivative function for zero values we find maximum and minimum points of our model. Practice applications of derivatives are numerous. Derivatives determine acceleration from speed; they determine rates of growth in economic models. Or they can determine the rate of drop in a fever patient's temperature as a function of the amount of aspirin in his bloodstream. What-If studies depend upon knowing sensitivity of goals with respect to different model parameters; derivatives are crucial to this.

An integral is a summing device. Its classical function is to find the area under a curve. In practical terms that means summing together all the positive and negative impacts of some parameter of your model.

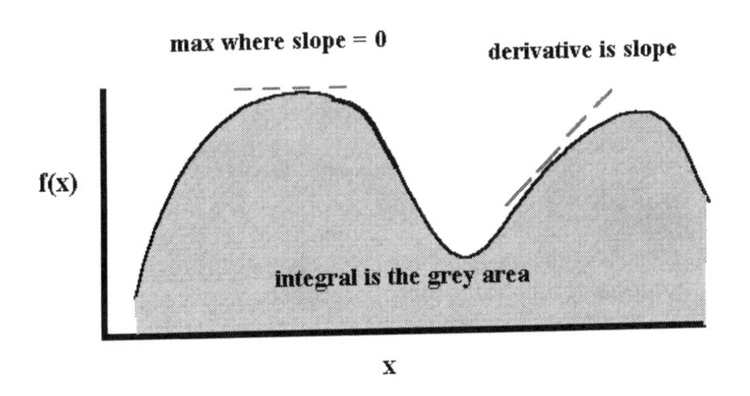

Calculus And Data Analysis

If you have a large data set that you wish to analyze, how can calculus help? Also, how much calculus do you need to know? Lastly, are there alternatives to a study of calculus? These three questions are what we focus on. Derivatives can be used for minimizing functions and for finding roots of equations. These two capabilities are interrelated. Since by max/min theory (from calculus) we know that a min or max is where the derivative is zero (or has a root). It's easiest to see this with an analogy to hiking. Suppose you are interested in hiking to the lowest spot in the surrounding area. One way to do this is to just start walking and change directions when a different direction is sloping downhill more than your current direction. Eventually you get to a place where all directions lead uphill. This is a (local) minimum, a valley. For most practical problems the approach is neither obvious nor easy. Let's see how calculus would handle problems like this. Recall the previous graph showing that the value of the derivative at each point was the slope of f(x). Since the slope line is going up from left to right by convention, it has positive slope. When you move in a positive x direction, you will increase the value of f(x). Conversely when you move in a negative x direction you will decrease the value of f(x). (These statements hold only for values close to x.) Since f(x) (in the graph) is always positive you will want to move in the direction of negative slope to decrease f(x). (The residuals function of our data analysis methods is always positive.) That, in brief, is the basic minimization approach for neural networks and nonlinear regression. You the user don't have to do anything. The program handles all the calculus automatically. A knowledgeable programmer already coded the calculus in.

A related calculus application is the case of linear regression. There are well-known formulas for finding the slope and intercept of a best fit linear equation to a set of data. Where do these formulas come from? From calculus that's where. Recall that the minimum or maximum is the place where the derivative equals zero. Statisticians originally found the linear regression formulas by choosing values for the slope and intercept that made the derivative of the residuals function equal to zero. Now you can simply use the formula and forget the calculus involved.

Engineering Modeling — Using First Principles And Calculus

This is the major league of modeling. This is where the models need to be exact; this is where the modeler needs a first class understanding of the principles of his area and a deep understanding of calculus. And yes, this is where the pay scale skyrockets. A calculus notion not taught until advanced calculus or later is that of a differential equation. This type of equation is far different from anything encountered in your standard algebra or trig classes. The solution methods can be incredibly complex. These equations are the core of most engineering modeling. If you have ever browsed an engineering textbook you have noticed that they write it in two languages: English and calculus. Page after page filled with derivatives, integrals and differential equations. To be an engineer you have to read and comprehend about ten to fifteen textbooks like this.

Usually, they derive the equations and theories in physics and engineering from calculus manipulations of a few basic first principles. Some of these are Newton's first and second laws, the principle of angular momentum, Kepler's laws, Hooke's law (elasticity), Bernoulli's principle (fluids), entropy, Ohm's law and Kirchoff's law (electricity) and Maxwell's equations. All are very important and interesting. Still, this whole area is outside the scope of this book which deals with modeling from data and analyzing data. We are *not* modeling from the first principles of science in this book. Most everyday problems can't be modeled well with differential equations anyway. They need algebraic models derived from database rules or statistical regression.

Checklist—Do You Really Need Calculus For Your Data Analysis?

So you don't need calculus at all? You can model without it? No, I am not saying that. There are many facets to the calculus-data analysis issue and there are many modeling methods. Let me try to summarize them.

Does Your Problem Really Require Calculus?

1. You do *not* need calculus for modeling with linear regression with a single independent variable. You might be able to get by without calculus for multiple linear regression. Still, it will be tricky if you have a large number of input variables or the inputs are correlated.
2. You do *not* need calculus for modeling with neural networks.
3. You *must have* calculus for large-scale nonlinear regression. Search-space issues like searching in the presence of local minimums and sensitivity to different parameters require calculus. For large problems these issues usually come up. (But even then neural networks can often do the same thing without calculus.)
4. You *can possibly do* small-scale nonlinear regression (one or two independent variables) without any calculus courses if you use one of the user-friendly solver programs available with spreadsheets. With a small number of independent variables you might be able to understand the search space problems and parameter sensitivity in an intuitive way and not need calculus. However, you could also be out of your depth. It is problem and user dependent.
5. You *absolutely must* have calculus and a lot of it for engineering style modeling.
6. In any case, your modeling will be *improved* if you understand calculus.

The message is that you can often get by without calculus. Your choice of methods will be restricted to algebraic models, with linear regression or neural networks. But that ain't shabby! However, knowledge of calculus will give you insights that can only improve your data fitting and analysis.

Chapter 8

Neural Networks For Beginners

Shortcut For Your Data Problems

Have you ever wished for a magical genie that would do your bidding? Or how about a magical black box that could give a solution to any of your problems? Well, for genies you are reading the wrong book—I can't help. As for magical black boxes that would solve your marital problems, your financial problems or the social problems of our country—again I can't help. However, for a mathematical black box that solves data analysis problems—I can help. Such a thing exists; they call it a neural network. Many real-world applications leave you a choice between some statistical method and neural networks. Until recently, you wouldn't have had any choice; only statistical methods like regression analysis would have solved your problem. Often to apply them you had to buckle down and learn some statistics. Unfortunately, if your math background was skimpy you might have to take prerequisite courses which include calculus. The amount of math required turned off a lot of people. These people had data analysis problems but didn't want to spend two or three years taking community college classes at night. So a lot of useful, profitable data analysis didn't get done.

That has changed. In recent years the rallying cry for the software industry has been "user friendly." For many folks that means "the less math the better." Complicated data analysis problems can be done with new neural network software packages. I think these programs are great and I recommend them to a wide set of users. This chapter discusses what a neural network (or mathematical black box) really is, how you can learn to run one quickly and the practical problems that you can do with one.

You don't have to understand much mathematics, computer science or biology to run neural networks. Jennette Lawrence, author and neural network expert explained it this way: "Contrary to popular belief, you do not need to understand the inner workings of a neural network before putting one to work for you." Commercial neural network programs take you through all the necessary steps without you having to be involved with esoteric math functions or computer theory. The program documentation is explanatory and usually includes examples on a CD-ROM that you can practice with and check your answers against. Still, you need to know a little about the general design process and the treatment of data. You need to be able to specify your problem with structure and precision; you must be able to specify a definite a goal, a predicted quantity. Most of your time will be spent doing that part of the job that you know best—choosing the information and data that will be used.

That's the beauty of neural networks: they do almost everything else for you. They only leave to you that part of the process that you know best—the specific information and data about your problem. You choose what might be relevant and pass it to the neural network. Notice the phrase "what might be relevant" in the previous sentence. Let's take time for a philosophical aside. They pattern neural networks after the human brain. Human brains are wonderful at drawing inferences from scattered observations, making analogies, identifying relationships, pattern recognition and generalization. Human brains are not so good at fast or large scale arithmetic operations, analyzing large amounts of data or many consecutive logic operations. Neural networks are not as good as humans at identifying relationships, pattern recognition and generalization. However, Neural networks are much better at fast, large scale arithmetic and analyzing large amounts of data.

Therefore, if you the human user of a neural network can get the neural network program off to a good start and correctly analyze its progress along the way, then you have the best of both worlds. If you, the user, can give good data to the neural network then the neural network can learn about your problem by identifying relationships between the input and output. That is really all it takes, data or information, believe me. Listen to Jennette Lawrence again: "Neural networks do not learn from a series of instructions or rules, such as cookbook recipes. From a network's point of view, there are no procedures; there is only information." Later you the user, can check your network's understanding by testing it—testing it by having it predict the output from data that it hasn't seen. This train-test phase is much like a teacher-student relationship. At the end you have a fully trained neural network that predicts the output of your problem. Here is where the teacher-student analogy stops—stops abruptly! After a teacher has trained a student all year and the student passes the final tests then the student moves on to the next grade. After you have trained a neural network and it passes the test data then you can use it for profit.

How does a neural network learn? Basically, it experiments with values of hidden parameters until its predictions match the observations. (This process will be clearer after the two examples shown later in the chapter.) Commercially available software is quite good at this and seldom requires any user intervention in this process. Your main job is to give the software a good start with a relevant set of data with well-chosen outputs. A neural network can learn relationships like, lower oil prices mean increases in stock prices for trucking companies, freezing rain sharply reduces sales (by keeping customers off the streets and away from stores) and a student with high 5th grade math grades and test scores will probably get high grades in 6th grade academic level math.

Just give it data? How about some rules-of-thumb or approximate formulas? Would these help too? Generally not. Data, good relevant data is sufficient. Furthermore, in many neural network programs you could not input a rule-of-thumb anyway. Usually, no mechanism exists for handling that type of information in a neural network. If you knew a good formula then it is possible to work that in. However, it would require either some transformation of the data or a software package with more advanced techniques. Data is what we want. The data should also include the output. Using our teacher-student analogy, the input data is like question to a student. The output data is the correct answer. So your data set needs many correct answers. It's not possible to tell you exactly what data you need, how much of it you need and what your output should be. Available data differs from problem to problem. Rather than go into general rules, we will cover a couple of practical examples first, then discuss generalizations later.

The Neural Network Method

There are several steps in the neural network method.

1) Define the problem
2) Gather the Information
3) Define the network
4) Train and test the network (this step can be repeated)
5) Predict with the network

The next two sections deal with practical applications of the neural network method; as we go through them refer back to this method. Our first example is a very practical situation involving sales and a perishable inventory.

The Neural Network Method—Bakery Inventory Case Study

Maria has spent her adult life working as a pastry chef. When a new mall opened close to her, one of the mall's committee members, who was a client of hers, approached her. The mall management team offered Maria a place in the fast food court section. Mall customers like fast food sections to snack at during extended shopping trips. Since keeping your customers in the mall is important to the retailers, a fast food court is essential to a mall's profit margin.

Maria didn't have any past management experience. In addition, the amount of space they offered her was only large enough for a counter, some tables, chairs and donut storage racks. She would have to do her baking before business hours at an adjacent restaurant. The restaurant didn't need its ovens in the morning because their main business was for lunch and dinner. So Maria started her store without a large capital outlay and her only employees were her and her mother. Still, it was great. No bosses, no hassles and all the profits were hers. Still, she has an enduring problem: her inventory is very short-lived, she throws out all the unsold donuts at the end of the day. Because she hasn't a kitchen of her own, she must do all the baking early in the day. That is a problem if sales are brisk because she can't go back and bake some more. No donuts late in the day mean dissatisfied customers and worse, complaints to the mall management who could retract her lease. So she always needs a cushion; this means some waste.

On the average she sells 1500 donuts a day. Depending on several factors like the weather and the day of the week sales have varied between 500 and 3500. Now if she always bakes 2500 donuts a day, then she is fairly safe; she will almost never run short during the day and lose sales, alienate customers, and risk the wrath of mall management. Still, her ingredients are the very best and she refuses to sell day-old donuts; that is how she got her reputation in the first place. The average cost of ingredients and baking (use of the restaurant ovens) is $.15 per donut: that is $150 of waste per day for a loss of more than $50,000 per year. But the alternative of baking 1500 donuts a day is worse. Not only does she still waste some on slack days; but often she runs short. Running short is a bigger problem: not only does it lose $.40 per donut of missed sales but customers complain to mall management. What to do is the issue. Maria feels that she can cut down and save money and still always have a cushion. To do this, she must accurately predict sales volume the day before and adjust her baking to match. She has observed several patterns over the last year: first, people often start diets at the start of a week, Monday, and for many resolve weakens by the later part of the week. Also, mall attendance is higher on Saturdays so the pattern is for more sales later in the week then at the beginning, other things being equal. Second, other things are not equal. Weather plays a part in both peoples' appetites and mall attendance. Colder weather does lead to more eating of donuts; rain especially freezing rain keeps mall customers away.

Seasons and especially seasonal holidays play a large role. Christmas season brings out the largest crowds. Summer brings teenagers who don't spend as much at other stores but usually don't count calories and are good bakery customers. Finally, there are holiday weekends like Memorial Day, Labor Day and the Fourth of July.

Maria reasons that if she could use these observations in some data prediction scheme then she could more accurately predict sales. A more accurate sales projection might cut down her 2500 donuts per day to perhaps 2000 some days, 1500 on others and save a lot of money on wasted ingredients. If she could cut down to only 200 wasted donuts per day then would be an extra $40,000 a year. Now her life has been better working on her own; but her income after expenses wasn't that much more than her salary in her previous job. An extra $40,000 per year would be great. Her mother lives with her and they could use an additional sun room in the back of the house. Also, they have talked for years of taking those guided bus tours of Italy and England, but there was never enough extra money. Less waste and $40,000 extra a year could do both.

Maria can handle a computer, but math is a different matter. She was good at algebra, even algebra II, but she never intended to go on to college so she never took the pre-calculus courses or beginning statistics. She knows her sales history for the past year and she can easily get the past year's weather data. Still, still she doesn't have any idea of how to use all that data to predict her sales one day in advance. What she needs is a black box that will do the prediction for her. A neural network can be that black box.

Define the Problem

What she needs is something that will take facts like: tomorrow is Thursday, during early November (a scale value of 1 for seasonal influence on sales) the predicted weather is for an average temperature of 42° with no rain. After processing these facts it would then predict sales of (say) 1650 donuts. Then Maria could add in a cushion for margin of error and bake perhaps 2000 donuts. This way she would save the $75 on ingredients that would be wasted if she baked that her normal 2500 donuts, 500 of which were just going to be thrown out anyway.

Gather the Information

Maria can put together a table for all of last year that looks like this:

Raw Data

Day Of Week	Temperature	Rain	Season	Donut Sales
M	40	0	1	1200
T	28	1	1	600
W	38	1	1	1300
Th	40	0	1	2100
F	46	0	1	2200
S	20	0	1	2400
Su	25	0	1	2000

This is what she needs. The two main observations are shown in the data columns. First, sales increase toward the end of the week all other things being equal (compare M and Th). Second, cold weather increases sales. Perhaps there is an important relationship between sales and these four data types; the neural network can find it.

Define the Network

Predicted donut sales are Maria's goal; so that should be the output of the neural network program. The other four data fields should all be used as input fields. So conceptually the neural network will look like the following diagram that we saw in an earlier chapter:

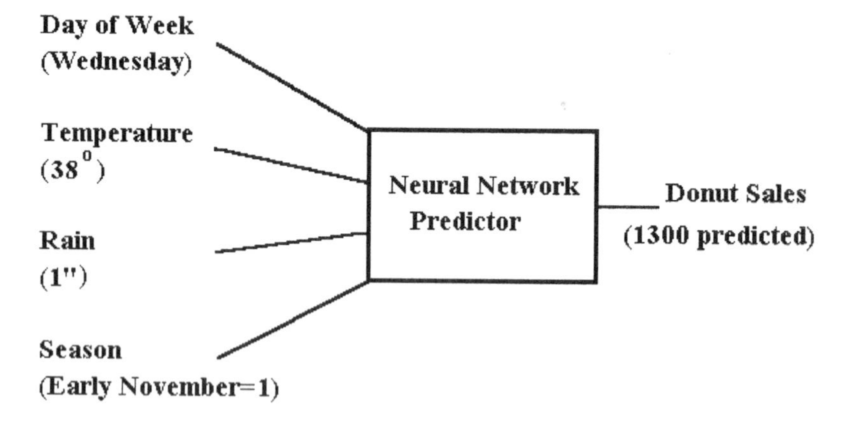

To set this up in most commercial neural network software programs you would first put your data in columns in a spreadsheet format. Then input them to the network program and label the day-of-week, temperature, rain and season columns as inputs and the donut sales as output. Note that all the data fields have to have numerical values. So giving the days of the week numerical values would be necessary (e.g., Monday=1, . . . Sunday=7).

The last thing Maria would need to do is to define how complex a relationship she wants the box to model. The number of hidden nodes or hidden variables controls complexity of neural networks. No perfect answer exists. However, choosing the wrong number of hidden nodes is not a fatal error. Maria (and you) will have many second chances. She can always rerun the program after the train-test phase with a more complex model (e.g., more hidden nodes). So she starts small and works up if necessary. One important hint: simpler is better, a simpler model that fits your data as well as a more complex model more often predicts better than the more complex model. You need to accept this point on faith since the point is difficult to show; it relates to over fitting data and there is a similar hint for general statistical modeling. For this problem a number of hidden nodes that is the average of the number of input and output nodes is a good place to start (2 in this case). Now she is ready to start her neural network program off and running.

Train and Test the Network

This is a good point to consider just how a computer program can take your data and automatically develop a prediction capability that exceeds your own. How does this all work? What is going on in that program? Maria declared four variables to be input variables, one to be the output and two hidden nodes in her network. Then the program took over and setup several equations and functions with several variable parameters. For example consider just one hidden node. As we saw earlier, the inputs feed into the hidden nodes where the mathematical model is developed and stored. One of these variable parameters of the mathematical model (call it" c_t", t for temperature) is a multiplier. The numerical value of temperature for each line of data is multiplied by c_t. Then this is added to the values of the other three data points multiplied by their respective multipliers. Finally, all this is added. This weighted sum is the total input that the hidden node operates upon. (For example, day=1, temperature=40, rain=0, season=0 so the weighted sum is: c_d *1 + c_t *40 + c_r* 0 + c_s* 0 where c_d (d for day), c_r (r for rain) and c_s (s for season) are the multipliers for the other three data types.)

The "c" parameters have values so this sum can be computed. Before starting, the network program will assign values to these parameters, sometimes even random values, the starting values do not really matter that much. What is important is that the main purpose of the program is to change and keep changing all these parameters until a good match to the output is found. It works like this: assume c_d starts with value 1 but when the network program computes the predicted output the value for the first line of data is 1700. This is too high. During the next cycle (called iteration) the value of c_d is changed to .8 (all other parameters are changed also). At the end of this iteration the predicted value for the sales are recalculated to be 2100, which is way too high. The network has moved in the wrong direction, but it now knows that and will compensate. At the next iteration the network program changes the value of c_d to 1.5. The new iteration gives a predicted value of donut sales for Monday to be 1100 which is much closer. By going through many iterations the hope is that eventually the network program will converge to an accurate estimate for most values of donut sales.

This change-parameter-values then compute-predicted-output-compare-and-change technique is the core of the neural network program. It is performed for all variables and each data type and it is done for thousands and thousands of iterations. Exact details of the process are quite mathematical. Backpropagation is the name of the process; it involves calculus partial derivatives and max/min theory. However, you the user, don't have to understand calculus or partial derivatives or even details of the process. It's all done inside the program. It is like a black box that you can't see inside. That is the beauty of neural networks.

Still it is good to see how this works. The next table lists the predicted output (predicted donut sales) for several iterations during this process. For comparison the actual data values are listed in the last column. You can see how the neural network predictions converge to the known output values.

Predicted Donut Sales
Converging to Actual Donut Sales

Iterations	#1	#100	#1000	*Actual*
	1900	1400	1300	*1200*
	2100	900	700	*600*
	2300	800	1400	*1300*
	700	1500	2000	*2100*
	1100	1900	2100	*2200*
	800	2000	2200	*2400*
	1300	2400	1800	2000

Study this table. On the line for each day the predicted value of the data gets closer and closer to the actual data. For example, the program originally predicts Monday sales of 1900 at iteration #1. Succeeding predictions move steadily down to 1400 at iteration #100 then to 1300 at iteration #1000 which is close to the actual figure of 1200. Similar results hold for the other rows. In effect the network model gets better and better as the program executes longer (more iterations). Eventually though, improvement stops. Notice if we stopped at iteration #1000 the model always predicted sales within 200 of actual sales. This is fine for Maria's needs. So Maria might decide that the model is good enough at this point. We say "might decide." Maria wants proof that the model is actually good within 200 or 300 donuts for *all* (or most) days not just those that she input to the model. There is the concern that the model will fit the parameters to this particular data and will not fit as well on other data. Conceivably if she put in a different time period of data then the model might not fit within 200 or 300 donuts. Then she could be in the same spot she is in right now. She needs an independent verification, but how?

Testing is the answer. You take the trained network (for this example it would be the network and hidden parameter values at iteration #1000) and you get an independent data set. Normally, you keep aside some of your original data to serve as an independent data set, so this step is no problem. Then you run the model against this independent data. However, now you don't allow the program to vary the parameters (neural network programs have a special testing mode just for this) to try to match this new data. Instead your program simply compares the predicted output vs. the actual donut sales value. You would set a criterion to decide if the prediction is good enough. A criterion might be something like: for 80% of the test data the model predicted within 200 donuts of the actual sales.

If the network model satisfies your criterion then you are done. If not, then you go back to the network definition phase; you need to change something to get an acceptable fit. You have several options. First, you can run the current model for more iterations and hope that it improves. Second, you can keep the same number of hidden nodes and let the program restart with another set of randomly chosen parameters. This often works. The quick technical explanation is that the old set of parameters might be found in a local minimum area of the network's search space. This new restart point might allow a better fit to be found. However, if you are non-technical, don't worry, try allowing the program to reset the parameters and have faith that this might work. Third, you can change the number of hidden nodes and restart the entire procedure with a different model.

Predict with the Network

Eventually, you should either get an acceptable fit or you decide that the data that you gave the network program is incapable of predicting the output you want to the accuracy that you want. Maria won't have that problem; any improvement is money in the bank for her and the neural network can clearly do better than her current method. At this point she uses the best fit neural network and each night she inputs the next day's predicted weather, the day of the week and the shopping season. Then she runs her neural network and it predicts tomorrow's donut sales. This prediction is a basis for how much Maria bakes the following day. She should bake slightly more than the prediction to allow for randomness (this is covered in the chapter on spreadsheets).

Stock Market Forecasting With A Neural Network

Probably more mathematical effort is spent predicting the outcome of financial markets than any other single human activity. And why not—it offers the promise of wealth, perhaps wealth beyond our dreams. For our second example of the neural network method we will try to design a network that will predict the market value one week from today. If this is possible (big If as future remarks make clear) then we could make a bundle trading in and out of the market.

Define the Problem

Although there are a variety of things you can buy and sell in the stock market let's restrict ourselves to one item: the S&P 500 index. This is a cross section of 500 stocks that closely parallel the entire market. (It is like the Dow Jones average which is for only industrial stocks and represents a much smaller number of companies.) You can also think of it as the value of an indexed stock fund. Because in effect, it is a fund with all 500 stocks represented. If the S&P 500 index value is at 1010 this week and 1024 next week then you could sell your shares and gain $14 per share. So that is our goal: predict next week's value of the index; if it is high enough we can buy at this week's price, wait a week, sell at a higher price and pocket the profit. Or better still we could buy options, more leverage, more profits. We want a neural network to do this prediction for us.

Gather the Information

Stock market data is plentiful; we won't fail for lack of data. Our problem will be choosing what to use; there is simply too much data to include everything. (Long aside, neural networks run slow even on the fastest PC. You won't have your answer back quickly. Speed depends on three things: 1) the complexity of the network which is based on the number of inputs and the number of hidden nodes, 2) the amount of data and 3) the desired accuracy. With 50 input parameters, a few thousand days of data, 30 or 40 hidden nodes and you have far, far too much computation to do. You couldn't do it all even if you left your PC running all week. In effect a problem this size, which is perhaps a very good stock market size, is too large; you must cut down something.) Now the Big IF: the stock market is so complex, so driven by small events, and so driven by what people believe not what is the "true" value of the stock that you probably cannot represent all that with a smaller number of parameters and hidden nodes.

(Another long aside: accurate prediction of the stock market may well be impossible—I just don't know. There are a lot of great minds out there in the world that believe that strictly technical analysis of the stock market is pointless. They believe this because of the overall randomness of the market. This means that so many people are running so many computer programs trying to predict the market that most of the investments are based on some computer projection. Then the net effect is that all the effort somehow cancels each other out. Remember you are buying from and selling to people that are running programs trying to predict next week's market too. They are your competitors. With all this negative stuff in mind, there is a positive side. Many technical analysts disagree; they have made money predicting the market trends and believe that you can too. And many have made a lot of money. You pay your money, you choose your theory.)

If all that sounded a little mealy mouthed and evasive remember that we are talking about spending very little effort, then turning on a computer and possibly getting rich. Maybe you can do it, maybe not. No guarantees. Anyway it's great fun to try.

Suppose you choose your inputs to be a few broad-based indicators of market performance. For example, yesterday's number of advancing and declining stocks on the NYSE and NASDAQ, the total volume of trade, current prime rate, current price of gold, the federal one-year T-bill rate and the 30-day rate too, and the trend in the S&P 500 for the past two weeks (this will be done with three separate data columns). Some of these data values are listed in the following table. There are too many data types to list them all in one row; so a small subset has been chosen to illustrate. Note that the first 5 columns are all inputs; only the last column on the right (in italics) is the output.

They call this type of data where a single data type is repeated several times on one line of data a "time series." Time series analysis is the subject of many books and graduate courses. This example differs from the donut-bakery example. It differs because we believe that patterns (dips, up or down trends, or staying steady) in the S&P 500 (and other market parameters) indicate trends in the market. We also believe that these trends can be used to predict future values. In the bakery example it was not important what last week's temperature or donut sales were. Every day is its own independent happening in the donut example. This is not so for the stock market. Placing several time-shifted values of the S&P 500 index in one line of data is how we include pattern data for the neural network to learn from. The last three input columns are all time-shifted S&P 500 data. The value in the rightmost column is what we would want to predict. This is the future value that we will not have when we use the network for actual prediction.

Date	Adv Issues	Prime Rate	S&P 500 -2 weeks	S&P 500 -1 weeks	S&P 500 current	*S&P 500 future*
1 week ago	1100	5.7	1016	1021	1019	*1027*
2 weeks ago	1400	5.7	1011	1016	1021	*1019*
3 weeks ago	1200	5.6	1009	1011	1016	*1021*

You should study this arrangement carefully; in effect it says that you believe that the recent pattern or market trend will influence next week's trading. Therefore, it will effect the value of the S&P 500 at the end of next week. This is almost universally believed; because market psychology drives trading and prices to a large extent. The rightmost column is the predicted value of the S&P 500 index for one week after the time of the line of data. In effect we are saying that we believe that the recent pattern or history of the index is also important to our prediction. Rising or falling or stagnant markets cause different market reactions in the future. Neural networks should be given this information.

We can enter data for the neural network to recognize patterns by taking the same data set and time shifting it. What you do is give the network the value for the week for which the data are collected. This data will include the S&P 500 value that day, the value one week before, the value 2 weeks before. This S&P 500 data will be used to predict the S&P 500 value one week after the week. Now your plan is to have the network predict one week in advance. Therefore, the network's predictions need to be measured against the value one week later. Then the network can work with years of historical market data and train by predicting the value one week after the data line that it uses as input.

Design the Network

Your table has five inputs and one output. As before, we start with a number of hidden nodes equal to the average of the number of inputs and outputs, three in this case.

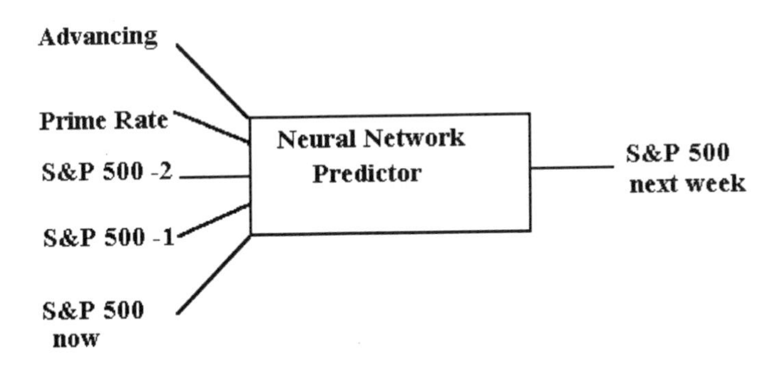

Now you are ready to start training and testing your network.

Train and Test the Network

This phase will work like you saw in the donut-baking example. There will be many variable parameters associated with the hidden nodes; the network software will vary them and compare the results with the actual S&P 500 index values for the next week. As the cycles or iterations go along parameters will vary and the predicted values will converge to the actual values. At some point the improvement will stop and you will have to decide what to do next. Consider the following table showing some results from different iterations.

Predicted S&P 500
Converging to
Actual S&P 500

Iteration #1	#100	#1000	Actual
710	1050	1030	*1027*
490	1015	1021	*1019*
1063	1030	1020	*1021*

At this point you would test this model against other stock market data that was set aside as an independent test. Then you would view the results and determine if your model performed well enough for you to invest your money on its predictions.

Predict with the Network

If you decide that the neural network is close enough then you would predict with it. You would do this by taking the input data at the end of a week. Then you would input this data into the finished neural network. Finally, you would run the neural network to predict the S&P 500 at the end of next week. With this prediction you would form your week's investment plan.

Advanced Stuff

Models and Weighted Sums

A magical black box this isn't. Although the high level diagram showed a box that made predictions, that box actually had structure to it and could "learn" the underlying math model.

A key point is this: to model a process it is *not* necessary to know the exact function form. Being able to approximate this exact function form with other functions is sufficient. This is what neural networks do. For example, consider a hypothetical system of two variables described exactly by the model $f(x,y) = \sin(x)*\cos(y)$. An approximate model that fits this equation is: $p(x,y) = x - 1/2xy^2 - 1/6x^3 + 1/12x^3y^2$ which is an excellent fit for small values of x and y. The approximate math model represented by the function p(x,y) bears little resemblance to the original model f(x,y); it is a polynomial function without any trig functions at all. Notice that p(x,y) is a *weighted sum of simple functions* like: x, xy^2, x^3 , and x^3y^2. Weighted sums of inputs are similar; a neural network computes a model that is like a weighted sum of input parameters. Neural networks do one additional computation besides the weighted sum. For technical reasons the sums at each hidden node are transformed with a function called a sigmoid function. Its graph is "S" shaped and it has several technical properties that make it ideal for extrapolating weighted sums of input parameters into a model. But that is all. A neural network is a weighted sum of input parameters. These weighted sums are transformed slightly by a sigmoid function. Then the results of the hidden nodes are summed together into other weighted sums. These weighted sums go either to a second layer of hidden nodes or to the output node, depending on the network definition. This may sound complicated but you can understand it with a little study. Of course, you, the user, don't have to understand this theory. Weighted sums of simple functions make excellent approximate models for most math functions and weighted sums of inputs make up neural network models.

What's in the Black Box?

The box itself actually is a maze of interconnections and nodes. The following diagram is the donut-bakery network. It has four input nodes, two hidden nodes and one output node. For this example all nodes of one layer are connected to all nodes on the next layer. Each of those connections represents one or two unknown coefficients.

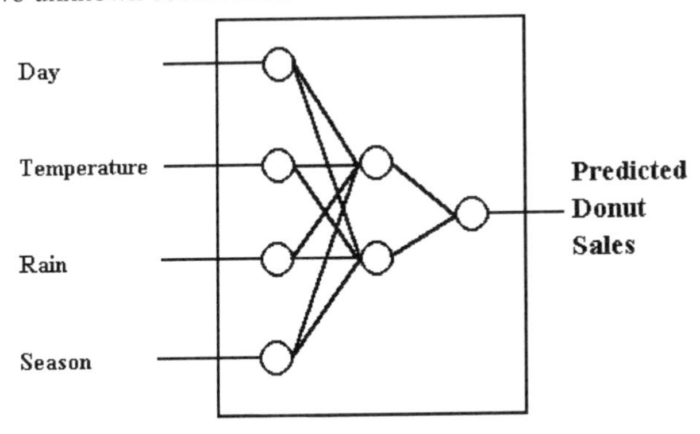

Notice that there are all three layers of nodes within the box: the input layer, the layer that we have been calling hidden, and the output layer.

How can this Work?

OK, you say, so we have weighted sums and nodes and connections and other transformations and the network iterates and converges to a solution. How does this thing do mathematics? The theory is very complicated but it is the same theory that governs statistical regression. A model is a mathematical function; usually it will be continuous. Mathematical theorems prove that weighted sums (linear combinations) of simple functions (polynomials or simple trig functions) can approximate any continuous function as closely as you need. The more accuracy you want, the more simple functions must be included in the weighted sum. This means that operations like division, multiplication, square roots, logarithms, powers, polynomials and more can be simulated inside a neural network by weighted sums.

Let's see how this works with some purely mathematical examples. Suppose you have just one input, a data set called x. Now if your output is the data set x^2 or the data set $3x^2 +2x + 1$ then the neural network can model it. However, if your output is the data set xy (where y is another independent variable) then the neural network cannot model it because you aren't giving it any information about y. It can't learn without the right information.

Try another example: let your inputs be xy, y^3, and y*sin(y). Now if your output is x then the neural network will model it. Something like this happens inside the network: first the neural network builds the capability to compute y from y^3 and possibly y*sin(y). This is done by computing the right values for some unknown coefficients. Next it builds the capability to divide xy by y by computing more values for other unknown coefficients. Then by dividing xy by y you get your desired output x.

The above example is more appropriate for real world applications because most data we get are composite type data like xy, y^3, and y*sin(y). We really don't know and can't measure a set of truly independent variables. For example, the price of silver is an input to the market example. Clearly, it is a function of many things like the Fed 1-year T-Bill, prime rate and even the S&P 500. The donut example is more like purely independent variables; the day of the week is not related to temperature or rain in any way.

Neural Networks run Slow

So far neural networks look ideal, don't they. They can take a wide range of inputs and automatically decide what operations to model in the hidden nodes then fit the data. They can do it all. In addition, they are easy for those that never took advanced math or forgot what they took. There is a downside to neural networks. They run slow.

A neural network training phase requires a lot of computer calculations. We looked briefly at the backpropogation method that looks for the set of values of the unknown coefficients that in turn minimize the residuals function. Technically, this is a nonlinear optimization method and these methods often need so many computer operations that large scale problems bog down even supercomputers. Large scale is key—if you can hold down the number of input data sets and the number of hidden nodes then your problem will probably remain manageable.

Here is a quick look at what happens to cause the trouble. Notice the previous diagram. Each input node is connected to each hidden node. Each hidden node is connected to each output node. For n inputs, m hidden nodes and p outputs you have *nm* + *mp* connections. Each connection has one unknown coefficient associated with it and there are a few extra unknown coefficients. The program must solve for each of them. A numerical example: for 10 inputs, 6 hidden nodes and 1 output node (a moderate size problem) you get more than 66 unknown coefficients. This will take some time to solve.

That is just the speed issue; another important issue is that of local minimums. A mathematical function with "nice" properties (continuous on a closed set) has an absolute minimum(s) which is the point that we seek for the "best" model.

This is the point where the residual function is at a minimum, where the model has the least possible deviation from the data. However, all known minimization methods get stuck on local minima (a local minimum is like a small valley with a depth of 100 feet that is miles away from the absolute minimum, the Grand Canyon). Often there are many local minima and this can slow the search for the best model.

In summary: your problem could take a lot of computer time, several weeks or more, to solve.

Checklist

The following checklist is designed to guide you through the major steps of a neural network modeling effort.

1. **Define Your Problem** Identify your goal/target/dependant variable(s) data sets; these are the ones that you wish to model or predict. These will be the dependant variables of the model and the output(s) of your neural network. Next identify your potential input/independent variables. Use your business area knowledge to get as complete a set as possible. This means that your independent variables can determine your dependant variables (e.g., a simple math example, if your dependant variable z is such that $z = xy^2$ then you need both x data and y data sets to determine the model, just x or y alone won't do).

2. **Gather Your Information** Gather your data together. Choose as uncorrelated a set of input parameters as possible. This means if two of your potential input parameter data sets are correlated strongly to each other then you should strive to use only one of them. Caveat: if possible, use your business judgement, this takes second priority to having a complete set of inputs from step 1. (Caution: this is important. It is critical in statistical modeling since if you put in two closely correlated or linearly related inputs your statistical matrix inversion subroutines can actually stop running! A divide-by-zero, a fatal run time error can occur. However, this problem won't stop a lot of neural networks since many of them don't depend on matrix inversion. People often see this as an advantage of neural networks. Yet the truth is that neural networks are dealing with the same search space as a statistical program and they can get hung up in local minimums too.)

   Split your data into two parts (if you don't have enough data to do this you should consider an alternative method to neural networks, the testing step is crucial for neural networks). The larger part will be the data that network learns with. The smaller part will be kept aside until the testing phase after the network is complete.

One last point: different methods are better for some problems. However, for neural networks the situation is different from that for statistical modeling, since neural networks usually handle multiple outputs quite easily. If you have a one-input parameter, one-output parameter problem, then I recommend that you don't use a neural network, use standard linear regression instead—it is simpler. In all other cases, multiple outputs and nonlinear problems neural networks will work.

3. **Define the Network** Complexity of neural networks is controlled by the number of hidden nodes or hidden variables. The number of hidden nodes in a standard backpropogation network is the analog of the order of the polynomial function in linear regression or the set of functions used in advanced multiple linear regression or the complexity of the nonlinear function used in nonlinear regression. Choosing the right number of hidden nodes is like choosing the right general form of the model. There is no perfect answer. However, choosing the wrong number of hidden nodes is not a fatal error. You can always rerun the program after the train-test phase with more or fewer hidden nodes. Simpler is better, a simpler model that fits your data as well as a more complex model more often predicts better than the more complex model. There are several schools of thought about where to start. Rough rules-of-thumb are available, one we have used in previous examples is to define the number of hidden nodes to be equal to the average of the number of input and output nodes. The documentation that comes with your neural network software will have more suggestions.

4. **Train and Test the Network** There are a few more parameters that need to be given to a neural network software system. These parameters relate to how the software will conduct the search for a best fit to the data, and how close a fit is acceptable. Your software documentation will have specific directions. At this point you are ready to run the network program. You start it and let it run. Walk away from your computer! As we saw earlier—neural networks run slow. It is not going to be done as fast as a spreadsheet calculation or linear regression. We may be talking hours or days here. So be patient, neural networks are not for those that want instantaneous results. Eventually the network will converge within your criterion or it won't

and you must manually stop it. If the network model satisfies your criterion then you are done. If not, then you go back to the network definition phase; you need to change something to get an acceptable fit. You have several options. First, you can run the current model for more iterations and hope that it improves. Second, you can keep the same number of hidden nodes and let the program restart with another set of randomly chosen parameters. This often works. Third, you can change the number of hidden nodes and restart the entire procedure with a different model; this is the equivalent of trying a more complex or less complex mathematical model. Eventually, you should either get an acceptable fit or you decide that the data that you gave the network program is incapable of predicting the output you want to the accuracy that you want.

5. **Predict with the Network** Now you are ready to use the results for prediction and profit. A raw neural network is a large table of long numbers; you need it in a different form. Your software will give you options: often you can simply provide a file of input parameters to the finished network and then run it in a different mode that will not fit this data but simply output the predicted output parameters. Another possibility is that the program will output a "C" routine that incorporates the neural network in its code and then you can add this code to your other software or a spreadsheet.

Neural networks and the regression technique overlap to a large extent. Either method can solve many problems. Neural networks are easier to use but slower; nonlinear regression requires more math of the user but both linear and nonlinear regression techniques run faster on a computer. My advice: if the problem is linear, definitely use linear regression. If the problem is nonlinear and if you have the math skills to use nonlinear regression then ask yourself the question: "Do I have a good idea of the underlying functional form of my model?". If the answer is yes use nonlinear regression. If the answer is no or if you don't have the math background then use neural networks.

Chapter 9

Basic Logic

For Those That Forgot Or Never Cared

"This stuff is all worthless. My data isn't tables of numbers. My data is text: words, policy statements, regulations, results of studies, comparisons, points and counter points. So much for all this math stuff, forget the trend analysis, skip the statistics and don't bother me with your neural networks. My data is words not numbers. So what can you do for me?"

Well, I am properly humbled. You have a different and indeed a tougher problem. Textual data is always open to interpretation. Even more, there aren't standard techniques like neural networks or statistics to do the work for you. Most all of mathematics can't be applied. Computers can't completely do the job either. Everyday millions of us have to analyze and derive results from reams of data. This data comes in all forms: government regulations, company personnel policies, business reports, sales progress updates, legal restrictions or foreign trade news and laws of all types. It's data too. It is all data. Our tools are not mathematical formulas, or statistical methods but our minds; our minds and their ability to read, interpret, follow logical reasoning and deduce facts. Logic is central to this; we can all read the words. However, they pay us for our abilities to interpret the meaning of regulations or for following the logical reasoning in reports or for deducing plans and actions. So it's with logic where must start.

I know what you are about to say: "No, not logic, it's the dullest field imaginable." You have a point. Still, the amount of logic that is necessary is quite small and we can go over it quickly. The main problem is that logic is abstract, with all of its hypotheses, conclusions, main premises, contradictions, fallacies, implications, paradoxes and the like. Some abstraction is necessary. However, we are all practical people here, more interested in real-world applications rather than theory. Let's start with a real-world business situation and then go on from there.

Katie has been a personnel officer at a large corporation for a couple of years. Her main function is to do the background work in finding relevant policy statements for different personnel issues that come up. In this regard she is like a legal assistant or a junior member of a law firm. She considers the issue, then pulls together the policies that might impact on the personnel case, next she puts together a tentative plan of action. Finally, she presents her case to a senior personnel manager for approval. This manager can accept, reject or modify her recommendations. Anyway, it behooves Katie to have a tight case. She must have all the relevant policies at hand so her manager doesn't think of something she missed. She should have a good logical defense of her plan, so the manager doesn't refute her logic. Finally, Katie must understand the implications of the personnel policies so she can answer the manager's questions about alternatives or modifications. Consider the following set of company policies concerning career development, career tracks, company priorities and training. This is the type of "data" Katie deals with daily. Not much of it is numerical, but her analysis and recommendations have to be based on logic and logic is a field of mathematics.

A The company list of priority jobs is: reengineering consultants, technical writers, electronic engineers and secretaries.

B To change jobs in the same field (same job title) you must have been on the job for 6 months and have received satisfactory performance reports.

C To change jobs to a new career field (new job title) you must have been at the company for 2 years, had satisfactory performance reports, and fulfill the certification requirements for the new job.

D To leave a priority list job (for another in the company) you need your supervisor's permission and an impact statement.

E To receive company-funded training in your current field you need your supervisor's permission.

F To receive company-funded training in a new field (priority) you need your supervisor's permission.

G To receive company-funded training in a new field (non-priority) you need the approval of the personnel committee.

H Situations not covered by any specific set of rules must be referred to the personnel committee for resolution.

Notice that this is a subset of eight regulations out of perhaps hundreds or thousands; Katie needs to use her own specific knowledge to pick out this subset. Notice also those regular people not necessarily lawyers or mathematicians wrote these words. The rules are not in a classical logical form like "*If* ... *then* ...". Finally, like all sets of regulations or laws there could be pairs of regulations that contradict each other and there are certainly omissions or gaps.

Logic Basics

Basic logic is simple enough when considered by itself. The applications of logic can be involved. Logic deals with statements or sentences that are either true or false. Other types of statements are outside the scope of logic. Examples show the difference: "Green is a good color for small sports cars" expresses an opinion but it doesn't have an intrinsic value of true or false, while a true statement is: "The sun rises in the East." Logic is a branch of mathematics where the variables like u, v, X and Y designate true or false statements, instead of numbers like algebra. In a sense mathematical logic is the algebra of statements. Numbers have several basic operations such as addition, multiplication, subtraction, division, powers and roots. When you add or multiply two numbers the result is another number. Logic is a little different. You can't add or multiply statements or sentences. Yet, there are operations that take two statements and combine them into another statement. The four basic operations are: AND, OR, NOT and IMPLIES. NOT operates on only one statement; it is simple negation: "The sun does not rise in the east" is a false statement, the operator NOT changes the value of the statement from true to false or from false to true. AND and OR operate on pairs of statements and give a new combined statement as the result. If two statements X and Y are combined into the single statement X AND Y then this combined statement is true only if both X and Y are true. If either or both are false then the combined statement is false. If two statements X and Y are combined into the single statement X OR Y then this combined statement is true only if either X is true or Y is true, the combined statement is false only if both X and Y are both false. OK, so what's the point of this?

They usually phrase rules and regulations as a series of logical conditions. For example: to qualify for a senior engineering position in Katie's company you need three things: a degree in engineering *and* six years of experience *and* performance reports of good or better. Notice the use of the word *and*: you must satisfy all three qualifications not just one. If the *and* were replaced by the word *or* then you would only have to satisfy one of the three conditions. Situations where there is a list of conditions one of which must be met are handled with the OR operator for this reason. Another business related example: you can qualify for reimbursement of college tuition if you are completing a degree program in a priority field *or* if you are working on an advanced degree in your field *or* if you get special permission from the personnel committee. You have options here; OR is the operator that joins options.

The last basic operator is the most difficult—the IMPLIES operator. It is distinguished by an *If ... Then ..* structure; if statement X is true then necessarily statement Y must be true also, This basic operation is trickier. Recall that statements can have either a true or false value. Implication is defined for all values of X and Y but it usually is applied only when the first term is true. That is: if X then Y says if X is true than Y must be true in normal usage; it doesn't say anything at all about the other possible cases. (Technical aside: you may have learned in math courses that the implication is true if X is false also. This is correct; the reason for this is that the logical operation implication when applied to two false statements must have a value. For technical reasons this value must be true. Although, in everyday usage it doesn't come up because nothing can be deduced from a false statement in a normal implication.)

One last logical entity is a logical function. Logical functions can vary in style and form. Nevertheless, they all have a value of true or false. Thus, they can all be treated as a logical statement and so combined into more complex logical expressions. Two examples from the business world: in Katie's personnel database there is a column for each employee's duration of employment with the company and for their current performance report. Therefore for each employee a statement like $duration \geq 6$ is either true or false and a statement like $performance \geq S$ (satisfactory) is true or false also. So these two items could be treated as logical statements and used in combination of logical statements.

That is enough basics. Now let's use them in real-world situations. Look at regulation B which is restated below and then look at it restated in implication form with extra parenthesis and ANDs and ORs.

B To change jobs in the same field (same job title) you must have been on the job for 6 months and have received satisfactory performance reports.

B' IF (you have been on the job for 6 months or more)
AND
(you have received satisfactory performance reports)
THEN
(you can change jobs in the same field)

The point to this is to show how they can rewrite a policy as a logical statement. Next, we need to see how they could rewrite this a couple more times into a format that is precise enough to be programmed on a computer.

Now we use logical functions that would be found in the employees' database and rewrite the policy statement using them. Compare and contrast rule B with its rewritten equivalent B". The former was written by humans with no thought given to computer processing. The latter is ready for computer processing by using fields from the existing database. The database software would recognize Statement B". The latter is also more precise. It is written with mathematical precision and can't be interpreted differently by different readers. At this point a good database program can handle these policy statements. We are going a couple of levels further into abstraction to show an example of how an AI program could do logical reasoning with these policy statements.

B" IF (duration ≥ 6) AND (performance ≥ S)
THEN (new_job_field = old_job_field)

Next, we need to introduce the standard mathematical symbols for the logical operations. They are: AND = ^, OR = v, NOT = ~ and IMPLIES = =>. Now rewrite the policy statement in terms of statements and their symbols.

B"' (duration ≥ 6) ^ (performance ≥ S) => (new_job_field = old_job_field)

Symbols simplify the concepts and the notation. By assigning sentences to symbols we can shorten the statement of a rule or regulation. For example: continuing on with our policy statement let X = (duration ≥ 6) and Y = (performance ≥ S) and Z = (new_job_field = old_job_field) then at last we can rewrite the statement in its most concise form:

B"" X ^ Y => Z

Logical Deduction

Definitions, operators, and symbols are the means to an end; our goal is improving our own logical reasoning and deduction. For Katie that means deducing facts, required actions and steps for planning from the list of policy statements. Katie is faced with the following situation: she needs a person knowledgeable in both electronics theory and signals analysis for a short term project in a newly formed engineering group in her company. These are both taught in college engineering programs but not all engineers have both of the courses and her database does not list individual courses that each person took. The database only lists college degrees and other certifications. During a phone conversation with the head of engineering she learns two facts: 1) signals analysis is a prerequisite for an advanced telecommunications certificate and 2) electronics theory is a prerequisite for signals analysis in most colleges. She does have a list of employees with advanced telecommunications certificates. From 1) and 2) she can logically deduce that if an employee "M" has an advanced telecommunications certificate than by 1) M also has passed signals analysis and thus by 2) M has passed electronics theory, too. Therefore M has passed courses in both signals analysis and electronics theory and is thereby a qualified candidate for this temporary reassignment.

Notice that the fact that an individual has taken and passed classes in these two areas is not listed in the database. This information had to be logically deduced from other information and implications. Although, a little tedious, it will be helpful if we go through Katie's logic step-by-step and symbolically. The reason for this is that later we want to look at how a computer expert system would do the problem. Expert systems can do logical reasoning with knowledge databases. People operate with words; computers must use symbols. So our example must be done both ways. Then we can see the connection and compare and contrast the human approach with the computer approach.

Back to Katie. She accesses her database and finds that an engineer named Tom is certified in advanced telecommunications. This is a true statement or fact; we can represent it as X. The implication: if you have an advanced telecommunications certificate then you have passed signals analysis, can be written as X => Y and the second implication that if you have taken signals analysis then you must have passed electronics theory can be written as Y => Z. Katie's logic would look like this symbolically: X, X => Y, Y => Z all are true. The pair of true statements X and X => Y together imply the new true statement Y. This together with the true statement Y => Z implies the new true statement Z. Finally, the pair Y and Z being true lead to another true statement Y ^ Z. This is Katie's desired conclusion. She was looking for a candidate that had passed both signals analysis and electronics theory and that is what Y ^ Z says.

Computers can do this too. Yes, computers can actually go through human type reasoning on data like this and draw valid logical conclusions. They call the field Knowledge Engineering or Expert Systems and it is a recent arrival in the computer world. They started most commercial applications like this during or after the 1980s. Very few real-world applications of expert systems predate 1980; also, the field is constantly changing. However, the basics of automated logical reasoning have stayed constant and can be explained quickly. As always, an example can best show the principles; afterwards, we can delve into the technique in more depth. Consider Katie's problem and the way she solved it. At each step she took two true statements or facts and merged them with a logical operation and derived another true statement. For instance, the first step was to take the two statements X and X=>Y and derive from them another new true fact Y. Then the pair of facts Y and Y=>Z were merged to derive another new true fact Z. Finally, true facts Y and Z were merged to form a true statement Y^Z which was the goal. All of this is listed in the next table.

Human Logical Process (3 steps)

Original facts or true statements: X, X=>Y, and Y=>Z

pairs of statements	derived fact
X, X=>Y	Y
Y, Y=>Z	Z
Y, Z	Y^Z done

A computer can do this too; however, it won't be as efficient since it doesn't have the human ability to pick out the best next step. Instead, a computer will have to loop through all the possibilities. The next table shows how.

Computer Logical Process (18 steps)

Original facts or true statements: X, X=>Y, Y=>Z

pairs of statements	derived fact
X, X=>Y	Y
X, Y=>Z	— (none)
X=>Y, Y=>Z	X=>Z

This was one iteration through the knowledge database of true facts and rules. During the processing two new true facts were derived: Y and X=>Z. These new facts now become part of the knowledge base and are considered in future iterations. Future iterations loop through all facts and rules and compare them two at a time. To keep from reconsidering two facts or rules that already have been considered together, the computer ignores repeats and only tries out new pairings.

Second iteration facts or true statements X, X=>Y, Y=>Z, Y, X=>Z

pairs of statements (those not done yet)	derived fact
X, Y	X^Y
X, X=>Z	Z
X=>Y, Y	—
X=>Y, X=>Z	—
Y=>Z, Y	Z
Y=>Z, X=>Z	—
Y, X=>Z	—

Third iteration facts and true statements: X, X=>Y, Y=>Z, Y, X=>Z, X^Y, Z

pairs of statements (those not done yet)	derived fact
X, X^Y	—
X, Z	X^Z

X=>Y, X^Y	—
X=>Y, Z	—
Y=>Z, X^Y	—
Y=>Z, Z	—
Y, X^Y	—
Y, Z	Y^Z
	done—the desired result

These symbols might make it hard to understand what is being done. Let's look at a couple of the intermediate steps and translate them back to reality to see what is happening. Look at the pair X=>Y, Y in the second iteration. The result was nothing—does that make sense? In English X=>Y means *a person is certified in advanced telecommunications implies they have take signals analysis* and the Y means *a person has taken signals analysis*. Clearly, nothing extra can be logically derived from these two sentences; so the result is correct. Another example: the first pair in the third iteration: X and X^Y means *a person is certified in advanced telecommunications and a person is certified in advanced telecommunications and they have taken signals analysis*. Again, nothing new was derived from these two statements.

Clearly, the computer method wasted a lot of processing on dead ends, pairs of statements from which nothing new could be derived. Also, the same facts were derived more than once, another waste. The total number of steps Katie, the human, took was 3; our computer needed 18, including many wasted dead ends or redundant facts. However, the computer is blazing fast compared to a human. Less obvious but actually more important: a human faced with a much larger database of facts could fail to see the shortcut method of deduction altogether. This human could end up doing as many operations as the computer; or worse yet never coming up with a logical derivation of the desired conclusion. This problem had 3 initial facts and the logical deduction process took 3 steps. What would happen with 20 initial facts and a logical deduction process that required 10 steps? Katie could easily miss the entire thread of logic necessary to verify the desired conclusion—I know I could miss it. Here lies the advantage of a computer. It essentially tries every possibility. If there is a solution it will find it.

Knowledge Engineering And Expert Systems

Computers can do human-type deduction and inference as we have just seen. However, this is seldom done. The human mind still does most logical inference of regulations, policies, laws and other textual data. Textual data differs from numerical data in one very important way: numerical data is ready (or almost ready) for statistical prediction just as it is. Tables of numeric data can easily be input to a computer program for analysis. Very little human intervention is required. Not so for textual data, the data is usually in a form like the policy statement B. It's only after a good deal of human intervention and work that it can be put in a computer compatible format like B''''. Usually, it's not worth the effort. The expense of translating all the data to machine format is more costly than having human analysts work the data directly. In addition, there are other problems: often logical contradictions are present in any body of laws, regulations or policies. The computer can't handle contradictory information. Humans must resolve it beforehand. Therefore, usually humans analyze textual data and computers analyze numerical data. That's the way the world is today. However, it may change. The AI revolution that started in the 80s is spreading and more expert systems are being designed to analyze, predict, monitor and control.

Knowledge engineering or expert system applications are usually handled by logic programming systems and logic programming languages. The programmer's job in designing a logic programming system is somewhat different from normal computer programming which is done in languages like Basic, C, C++ or Java. A logic programming language, like Prolog, gives the programmer the ability to quickly input the data in a form similar to B''' or B''''. When the program is run the language then automatically goes through the logical deduction process and determines the solution. The computer is used as a high speed inferential engine which rapidly deduces new facts from the input data and strives to prove or disprove your problem. Expert systems have the advantage over humans in several ways: first, computers don't make mistakes on their own, a human might make a logical deduction error and conclude something that is false. A computer won't. Second, a human could lose their way in a maze of rules and regulations. When many different rules might apply in different order the possible chains of logical deduction grow rapidly and a human might miss the quick, easy solution. A computer will do the work by brute force at high speed. A maze of data will take the computer longer—but the computer will not get lost or confused. Third, the computer will probably be faster for larger data sets. If a rapid solution is important, a computer is the way to go. This is often the case: many real time control systems: air traffic control, medical monitoring devices, auto-pilots on aircraft, telephone and Internet load handling and many military command and control processes must be done rapidly to be successful. Too rapidly for humans it turns out. Instead, humans program the necessary logic into an expert system. Then the computer runs the logic against the incoming real time data to produce real time decisions that no human could do. Most of these applications involve more numerical data than textual but the computer has the same speed advantage for textual manipulation too.

Logic programming systems come in a variety: there are those based on lower level language code like C, C++ or Pascal, there are many Prolog-based systems and there are some business-area specific commercial products available (most are for hybrid systems for business systems analysis or financial investments). If you want to try to develop a logic-based system on your own, I recommend learning Prolog. As computer languages go, it is no harder than most. However, some people have trouble with the underlying theory of symbolic logic. If you can already program but never use any language but C or C++ you could look for logic programming libraries in C or C++ either commercially available or on the Internet.

Often AI software is hybrid; many applications require both logical data and numerical data. Although this book has a separate treatment for each type of data, you can combine expert systems with linear regression or neural networks for a hybrid system. And last and most important: the main skill you need is your own logical deductive ability. An expert system is a reflection of its programmer's understanding; expert systems are faster and more thorough than humans, but they can't perform any better than they have been programmed.

Should you use an expert system for your application? Tough question—often the answer is no. The main reason is that cost-benefit issue we referred to earlier. A lot of work goes into textual data preparation for logic programming. For even moderate sized company policies this could be many person-years of effort. Text consists of words; words have several meanings. Words must be interpreted in context. This is an entirely human activity. It is slow, it is costly. In addition, your experts that interpret the text need to be knowledgeable of symbolic logic and programming or you need a programmer that is also knowledgeable about your business area. This is seldom the case. All of that said, suppose you are expert in some area and feel that you could by yourself do the translation of this expertise to computer compatible logic. Also, if you are willing to learn how to use a logic programming system—then try it. For a long time you will be testing your system against your own knowledge and deductive ability and you will win, you will make the better judgement. Your computer isn't dumb. It just needs more information and better rules. Then you will add in more rules and data to improve your system. There will come a day when you and your computer disagree and after reconsidering everything you realize that the computer is right. At this time you have done well on your expert system and your work has paid off—now the computer can teach you new facts about your own field. In a way your children have grown up and now you can now learn from them. There is no feeling quite like it.

Devious Deductions And Lowdown Lies

After statistics, no area of math is as much abused as logic. Illogical conclusions, deductions with no rational basis and utterly baffling arguments are all commonplace. They have written entire books about the misuse of logic and numbers in everyday life. It's not our goal to discuss all possible abuses. However, a good understanding of the areas of a logical argument that are abused is very useful. This chapter summarizes most common logical problems in a concise appeal to logic basics: the logical implication, logical axiom systems, logical facts and others.

There are many logical fallacies. One good book on the subject is *The Power of Logical Thinking* by Marilyn Vos Savant (you know her: author, the columnist—*Ask Marilyn* in Sunday newspapers, highest recorded IQ in the *Guinness Book of Records*—228). In one nine-page section she lists no fewer than 22 different fallacies. All of these have names, most in Latin (Marilyn's Latin is much better than mine) and they can be grouped together to some extent. My approach will be that of a computer programmer and mathematician (well that is what I am). In this approach we will look at the basic patterns of logical deductions. Then we see how we can go wrong by doing something illogical to either the premise or conclusion or the axiom system or the interpretation of the facts. This will give you the basic framework for understanding fallacies if not a complete classification system with names.

An ideal and correct logical deduction normally will have the symbolic form like one of the following: A and A=>B are true or A and A=>B and B=>C are true. Here A is a true fact and A=>B and B=>C are true rules. The first case is straightforward: we know the premise and rule hence B is also a true fact. In the second case we deduct B to be true first then apply the second rule to infer the truth of statement C. We have then deduced two new facts B and C. Usually, C is the conclusion we are after and B is simply an intermediate step. This is the way computers do logic; it is the logic of Aristotle and mathematicians. It is *not* everyday logic. Issues by their very nature have two or more sides. If an issue could be resolved by an obvious chain of logic that was agreed upon by all concerned—then it would not be an issue—it would be an accepted fact. Usually "logical deduction" in real life is more loosely based. Perhaps the deductive process might look like this: A' is true (A' is some statement "close" to A), A=>B is true and the speaker wants you to believe B. Hence, the logical argument is presented: *A' is close to A* and *A=>B* is true therefore B is true. Purists would say that "close" only counts in "horseshoes and hand grenades" and this reasoning is simply wrong. Issue advocates would say that the purists are being too strict and that their argument is true enough. Which side is right? Neither and both. "Close" is a judgement call; we simply can't tell how close is close enough in advance.

Let's look at several examples of "close" variations of statements and rules. Rule: Taking one sleeping pill helps you sleep. Symbolic form: B = taking one sleeping pill, C = helps you sleep, B=>C). Try some other B' statements that are similar to B perhaps even close to B and see what happens. Say B' = take one pill of any kind; this is unlikely to help you sleep if the pill is an aspirin or antibiotic. The logic falls apart in this case. Rule: B' = take 2 sleeping pills. This might not be medically advisable but the conclusion is probably true. But rule: B' = take 50 sleeping pills doesn't work logically or medically, you don't sleep better, you die. Try a political policy: Adding 1000 new cops to a city's police force will lower street crime. Here B = 1000 new cops, C = lower crime rate. Let B' = 10 new cops. Can we still count on a decrease in crime, if only a fraction? Not necessarily, this is a situation that has come up in several cities. Some background: lots of police forces are reactive, police generally stay in cruisers or at the precinct houses and respond to 911 calls. Some police forces are more proactive and have pairs of cops walking the streets, a visible presence and deterrent to street crime. However, 911 calls have priority; so adding a few extra cops (say 10) to a large city force won't change the way it does business. However, add a large number (1000) and now the police can do both: respond to emergencies and maintain a presence on the streets. This helps, sometimes a lot; several cities have gone to this method. Back to logic: B and B' differ in substance not just relative size. What happens is that there is actually an intermediate step D = send foot patrols out to trouble areas. This is what lowers crime. In symbols B=>D and D=>C but B' doesn't imply D so it doesn't imply C (necessarily). The key point: it is not at all clear from just a statement of some facts and the form of the logical deduction that this is so. You need to know quite a bit of supporting information. The extra information you don't know, the extra information the issue advocate doesn't tell you, the extra information a politician avoids mentioning—it is this extra information that you need to understand most social issues and their logical consequences. Strict mathematical logic is good to know because you become aware of abuses; however, the real world doesn't work (even stronger: usually can't work) in the strict logically correct way. Strict logic seldom works in political discussions.

All the preceding dealt with fallacies involving the general form of the argument. Another source of fallacies is the set of rules/facts accepted as true originally. In math terms they call these assumptions, axioms or theorems. You can think of them as simply rules or facts that they accept as true. In this country we have many sources of truth: the Bill of Rights (Freedom of Speech, Right to bear arms, etc.), other constitutional rights (due process), civil rights, human rights of various types, biblical teachings, political correctness and others. Political discourse often involves appeals to different authoritative sources. Political discussions or arguments quickly break down for lack of agreement on the basic assumptions or axioms. Just think of the national debate on the death penalty, abortion, the military budget, gay rights, to name a few. How often have you heard a debate on these issues quickly change to two people talking *at* each other instead of talking *to* each other? Usually the issue is their core beliefs. The death penalty is a good example. Many people consider it a fundamental axiom that a legal system has no right to take the life of a prisoner under any circumstances. Many others feel so strongly about an individual taking the life of another that they feel that the ultimate penalty is not only justified, but required. There is no way around this. These two groups each have a different set of core beliefs that logically lead them to opposite conclusions. You can't rant and rave and insist that a political opponent should behave logically. They don't have to believe your assumptions/core beliefs and they won't.

Checklist For Analyzing Logical Arguments

We can look at examples forever; however, we are people in a hurry, people that want to fill gaps in our understanding of mathematical methods that effect our working lives. Instead of a thousand examples, a checklist.

1. **Overview of Argument** Look at the entire logical argument. Divide it up into three parts: logical structure, basic assumptions or accepted facts (include implicit ones), and data.

2. **Logical Structure** Isolate implications or logical deductions. Diagram the argument with letters like B and B=>C. Step back and analyze the logic. Is it rigorously logical? If so you are done. If not, say you have a B' and B=>C type argument; then you need to see how the argument varies from strictly Euclidean logic. Does the structure give you any hints as what statements or assumptions to concentrate on? How different is B' from B? What could go wrong? What else might you need to know?

3. **Basic Assumptions** If possible isolate the basic assumptions or axioms that the speaker considers true. Can you identify any apparent agenda? Any biases? Do you know enough about the subject area to evaluate the premises and understand the argument?

4. **Data** What is the source of the data? What, if any, statistical terms or processes are used (e.g., average, mode, trend). Did the source or the speaker interpret the data?

This checklist will give you a good start. Any obvious misuse of logical structure won't fool you. You will have some idea if you are listening to a speaker with a hidden agenda and you will be able to initially evaluate your knowledge of the subject area. After that you are on your own. All logical traps can't be categorized. Logical structure—there are only a few structural misuses; by diagramming the structure at least you will understand that part. After that the knowledge that you need is hidden from you. Data—if the speaker didn't volunteer the relevant data and you don't know it from your own experience, then you have a problem. Basic assumptions—a speaker's core beliefs are often hidden inside his mind, you don't know and may never know exactly what the speaker believes. His argument may seem like perfect logic to him and you won't have a clue.

What if you are an average worker in a large corporation and want to be an expert at all forms of logical discourse? Unfortunately, there are only two truly excellent sources of formal logical training: advanced college math and law school. I don't recommend either. However, reading books about logical abuses will make you aware of subtle arguments. Finally, some good active training is available to everyone by just reading a LSAT preparation book and doing the sample test questions. Good luck.

Chapter 10

Mining Databases

Structuring Your Data

Have you recently ordered something from a catalogue over the phone? If so you first gave the catalogue number for each item. Then the operator told you if all items were available, confirmed the shipping method and gave you the total cost. Then you read your credit card number to the operator and it was confirmed. After giving your name, mailing address, catalogue code number and a daytime phone you said goodbye and the transaction was complete. A few days later your items arrived in the mail. This was a database system at work.

In the background was a computerized database system. It recognized items by catalogue number, then reserved them for you. It entered your address for a mailing label, then it set up a message which informed human operators what items to package together and mail. It printed out the mailing label and automatically debited your credit card account. Very little human interaction was necessary; a modern computer database system did most of this.

Information is everywhere and so are databases. The ability to store, access, change, update and analyze data requires most adult workers to come in contact with one or more computerized database systems. You have to know something about these systems in the computer age. Luckily, database systems are usually easy to use at the basic levels. Still to be successful, to use your company databases or to use databases for your own business—you need to know more about database systems. This chapter aims to do that.

There are a few basic concepts that need discussing; a real-world example is the best place to start. Cynthia's middle school has data, a lot of data. Student names, addresses, home phone numbers, parents' names, age, sex, ethnic group, medical problems, school bus assignments, individual student class and room schedules and many more. Two key points: different people enter this data into a computer and different people are allowed access to different parts of the data. Nothing surprising here: parents fill out preliminary forms on their children with information like name, address, and age, and school administrators fill in additional information about bus assignments and home rooms, school nurses might note individual visits to their office and the medical problem, principals might enter confidential information about social or disciplinary problems and each teacher enters thousands of scores, grades and evaluations. For confidentiality reasons, not everyone can access all parts of the database. Certainly, no student or parent has any business accessing another student's information. A bus driver doesn't have access to medical information or grades and neither should a janitor. However, the bus driver should have a list of students assigned to her bus. A principal will probably have access to everything; but even she would only have the right to read all the data; she shouldn't have write access, the ability to change medical records or grades. Computer database systems need to keep all this straight. Different data from different people must be combined logically and data must be only accessible by those with a right to see it.

Those two requirements lead to a design of database systems. They call the most common form of a database a relational database. It refers to data stored in different tables. The columns in a table each have an identifier (e.g., name, address, sex) called a *field*. They call individual rows of data (usually referring to all aspects of one student or bus or class) *records*. For example: an individual student record would contain the student's name in the "name" field, the student's age in the "age" field and a set of similar records would form a table or relation. This table would be one of many tables that form the entire middle school database.

Different people would enter data in different tables. Different people can access different tables. Tables form a logical decomposition of the entire database. Designing a database is a technical endeavor and good design technique hinges of the proper choice of tables and fields. Therefore, you might think that understanding tables and database design is central to success with databases. That usually is not true.

We are at a fork in the road. Roughly speaking, there are database designers and there are database users. Database users far outnumber database designers. One person can design a database for a thousand users. Database design is perhaps a 6 or 7 on our difficulty scale whereas database usage is about a 2 or 3. You need to know where you fit in. If you work in a large company with a dedicated information systems staff then you are most likely a user. Here, knowing database design jargon and techniques is nice but not essential. However, if you are self-employed and have to set up your own database from different data types then you must learn something about database design. Here though, you won't need to design as complex a database as the one in a large company. You won't have to worry about who should have read or write access to tables of data. You are the only user; you get access to everything. A much simpler design is possible; a difficulty level of 3 or 4 is about right for this case.

The three main parts of database systems are the database itself, queries and reports. Usually, a user has a question (query) or a set of questions about the data within the database. After the queries are run against the database, the answers are formatted into a report. Queries come in many forms: you could query the middle school database to print out a list of all students that ride bus 12A. Or you could query the database for a list of all sixth graders' averages in math class and plot the results on a bar chart. A simple query might be to find the home phone number for a specific student. A report in these cases would be a list of names, a bar chart or a single telephone number. For most users a knowledge of queries is enough to intelligently use a database system. Therefore, that is what we will concentrate on: queries, what they are, what they aren't and how they work.

Finding Information In A Database - Queries

Just how does a computer find a specific piece of information in a large database? Since this is the fundamental operation in database systems, we must spend some time considering it. Our databases are in table format; a query might ask to find all records where the "class" field represents a sixth or seventh grader. All class data would be stored in a table under a column header "class," the class field; the computer would go to that column and then go through every table entry in that column. These would all be numeric data, either a 6, 7 or 8, and a logic comparison of the form: *data < 8*? would determine if the record was that of a sixth or seventh grader. Or search for a child named Johnny Smith, here the computer would go through data in the name field column of the table. A logic comparison of the text strings of this form: *name = Smith, John*? would be performed and the search would stop and return the entire record and the result "yes" if the record was found. Stepping through a table one line at a time is time consuming even for a computer; a faster method is a dictionary type search (called a binary search in computer jargon). Here the data are sorted beforehand in alphabetical order. The computer starts at the midpoint of the table and then moves up or down depending on how close the chosen entry is to the desired "Smith." For example the first name chosen might start with the middle letter of the alphabet M. So the computer does a test and finds that first letter in Smith S is after M. The computer's next step is to go halfway between M and the end of the table, names starting in Z. Now say the halfway point is a name that starts in the letter R. Now since S is still greater that R (greater as in alphabetical order) the next step is to go halfway between R and Z say V. Oops too far. So the next choice is halfway between R and V and this process continues after you get to the S names. Then by starting with the second letters and continuing until you have a complete match to the name Smith. (Aside for the math inclined: the number of items searched in an N element table is a small multiple of $\log_2 N$ by this method. For large N, say N = 1,000,000 this log term is only 20. So binary or dictionary search compared to searching every item is about 20:1000000 or 50,000 to 1—an enormous speedup. This is why most important fields in a database are ordered—"indexed" in computer jargon. That is computer search in brief; a computer going through many records performing a simple logical operation for each. The basic idea is simple enough; the need to sort or index is also easy to understand.

A query is stated in terms of logical operators like: AND, OR and logical functions like: equals, greater than or less than. A query must also refer to one or more fields within a record. The result of the query is a subset of the table(s) that the query is applied to and for which the value of the query is true. Most commercial software database systems try to be user-friendly; to this end they allow very English-like queries to be stated. For example in the mail-order database a query of the form: "Find all names with 'Mr.' but not 'Mrs.' in the title." This is simply English but it also is a logical statement of the form: *(title = Mr.)^(~title = Mrs.)*. The software can translate for you; still you need to remember that you can only query a database with statements that can be reduced to a logical expression in terms of field names that are present in the database. Database systems, currently, do not have many AI capabilities included. Hence, users can ask the database to find all records with Mr. but not Mrs. in the title or to find all records with the abbreviation "Apt" in the address field. These questions are correct queries. However, many questions cannot be reformulated in terms of logical expressions with these database fields—these questions are not correct queries and will not work. Examples of seemingly legitimate questions are: "find all addresses with bird names as street names" or "find all customers with Irish surnames." There is a problem. Unlike queries that match a known text string like Mr. or Mrs. the text string "bird names" and "Irish surnames" are not specified. The computer doesn't know what letter combinations form bird names or show Irish ancestry. A human could manually sort this database with these conditions. Nevertheless, the human is actually drawing outside knowledge from other databases inside his head; databases about birds and surnames, not just the mail-order database. We did not give our computer these extra databases; how is it to know? It doesn't and it can't. It needs the extra data to process such user requests. Queries must be stated in terms of logical functions and operators acting on database fields and values that are within the database.

The mail order example included information about prices of houses and apartment rents. These were some rough indicators of individual household income. These data were not part of the original mailing list. Recall, you can obtain this data from publicly available data. Still, it represents a different database. This database must be included along with the mailing list. The mailing list database might have these fields: name, street address, city zip code. The real estate value database might have these fields: street name, range of street numbers, rent (if an apartment), value/price (if a house). Some explanation: not every house is up for sale so you can't get an assessment of individual homes easily, but you could input an average value for all homes in a given development, then put in the street names in that development and the range of street numbers corresponding to houses within that development. You might end up with one number like $170,000 representing the value of each of 500 homes. It is a rough estimate, homes could vary by $30,000 or more, and actual household income, which is really the figure you want to estimate, could vary even more. Still it is an estimate and will help you refine your mailing list.

Those two tables must be combined to get an estimate of housing price or rental cost for each address on your mailing list. Look at the two sets of fields: mailing list: name, street address, city, zip and real estate table: street name, range of street numbers, price, rent. From the mailing list you can split each address up into a street number and a street name. Now the street name is a common field with the real estate table. This street name field links the two tables together. A record from the mailing list can be cross referenced to the real estate table via the street name field. Then you know an estimate of the house value or rent at the address.

This linkage of field names that are common to two or more tables is the primary method of making complex inferences from large databases. Another example based on the middle school database is that of student IDs, or personal identification numbers (e.g., social security numbers). These appear in several middle school database tables. They link the tables together; they are unique student identifiers. If you access a school bus record and find the names and the student IDs of all the passengers then you can use these IDs to access another table to find the home phone numbers of all the students. A school nurse might want to call all the parents of the kids on one bus if they were all exposed to a very sick and infectious bus driver. Linkage between data tables is essential for that. Now the good news: database software does linkages automatically. A user doesn't ever have to know what table the data are stored in; he can query the system and it will respond by gathering the data from all the pertinent tables. The essential item is the field name that is unique and links tables. Usually this is a database design issue; we call these fields *key fields* or *keys* and the design of tables and relational database systems hinge upon them. However, it is a design issue and if you are just a user you won't have to concern yourself with these issues. All that a user has to know is how to query a database, leaving the rest up to the computer.

Databases have several math and logic functions available to manipulate the data. We have already seen logical functions like = or <. Other math functions are average, sum, scale (multiply by a common factor), divide, absolute value, some trig, exponential and logarithmic functions, plus a few others. A major function provided by databases is the sorting operation. We humans like our world ordered or sorted (indexed); we want to know where to look. Also, as we have seen binary search allows much more rapid search of a database. Often we sort or index databases on several fields, like names in alphabetical order, street numbers in sequential order, or addresses grouped by zip codes. Sorting is probably the biggest single database operation that you will use.

Conclusions

To repeat, database usage is easy, 2 or 3 on our difficulty scale. However, large database design is more difficult, 6 or 7 on our difficulty scale; however, few people are actual large system designers. If you are employed by a large business and use preexisting databases, you don't have to know that much. A weekend of practice with the computer software and its tutorial should be enough for 90% of all users to get started. If you own a business and want to set up a single user, moderate-difficulty database then you have an intermediate level problem, 3 or 4 on our difficulty scale. A two or three-day course at a local computer school would be appropriate. However, if you have a simple database requirement like the mail order example then again self study for a weekend should be enough.

Spreadsheets or databases: which should you use? Often it is not clear cut. These two systems have a lot of overlap in capabilities and as both have evolved they have added more of the other's capabilities. They are growing closer together. For simple tables both will work. Spreadsheets can do basic database operations like searching and sorting; databases can do basic spreadsheet calculations; both spreadsheets and databases offer good reporting capability. I've taught classes for years and for the straightforward process of grading, I have used both spreadsheets and databases, each about half the time. Training is available for both: great books on specific systems are available at all large bookstores; on-line tutorials are excellent and just playing around with sample database or spreadsheet examples will teach you a lot.

Databases are as easy as anything in this book. They are universally used; knowledge of them will help at work and in a home business. Good high quality instruction is readily available. If you have never used one, don't be the least bit concerned: jump right in.

Chapter 11

Lost In A Large Corporation

Data Analysis For Non-MBAs, Non-Managers, Non-Statisticians And Other Regular People

Maybe you are like millions of other Americans: you work a forty-hour week in a medium-to-large business. You enjoy your work; however, you want to do a better job. You want to impress your coworkers and your boss and you want to influence the direction your company takes. You aren't upper management, you aren't one of the top planners, you don't have that magical MBA. You feel like a small-to-medium sized fish in a medium-to-large sized pond. Yes, you have data, but no, you don't need a predictive model—others in the company do that. Can this book's program help you?

Maybe you want to help reengineer your business processes. Maybe you want to be a member of a planning or steering committee. You are a recognized expert in your own business area: maybe personnel, training, engineering or production. Still, you feel too narrow, that your ideas don't really effect company-level goals. Or if they do effect them you can't articulate how. How can you leverage your area-expertise into positive recommendations that effect company-level goals? Can this book's program help you?

Yes, it can. The result of modeling with data doesn't always have to be a numerical prediction or an optimal solution. Often, actually most often, data analysis and models lead to incremental improvement in business practices. Yes it can help you connect your area-expertise and data to larger overarching company policies. Doing it won't be easy. Most of the difficulty won't be the mathematics. It will be you making logical connections between your understanding of your business area and the larger company goals. The math part is actually quite easy.

Q I've read the book so far. I've seen the examples and checklists. But models are functions with independent and dependant variables, correlations have cause and effect parameters. Our company has thousands of data sets. It's not so easy to pick out the right data as it is in the bakery or mail-order business. Where do I start?

A Basically you start like this: First, view your business area and its data as the causes of the company goals data sets which are viewed as the effects. Second, company goal data are often represented by bottom line concerns like: profit margin, high production or low employee turnover. Try to identify a few data sets like this. Third, your data, the data that you understand best, might be data about individual employees: salary, duration of time with the company, amount/type of training, overtime worked, conferences attended, number of complaints filed, etc.

Q OK, I see that. I guess that it makes sense, trying to see how my business area effects my company's bottom line. But exactly what can I do with this data and how can I turn data and math into solid business decisions?

A Start by calculating correlations between your area data and the company goals data. This will yield interesting results and start you thinking about business process improvement. Maybe the employees' turnover correlates positively to employee performance reports. You, the personnel expert, interpret that to mean: good employees are leaving the company relatively faster than poor or average employees. Obviously, not good, but why? Maybe 1) the company pay scale is too flat and good employees are getting substantial raises to leave and work for other companies—do more correlations with the salary data. Or 2) maybe there is too much stress put on the highest performers, they are overworked—do more correlations with the overtime data. With some logic like this and some more correlations you decide that the data shows that your company's best employees are leaving relatively faster. The reason is that the high end of the pay scale is too low to retain them. You have some correlation evidence to support this. Write a one-page memo. Send it to your boss or present it in a weekly staff meeting or if you are shy put it in the suggestion box.

Q Sounds good in theory but my company won't do anything. I don't have any planning or management experience and I am not a mathematician. They won't change the pay scale based upon my analysis. Why should they?

A I agree. They won't, at least based solely upon your analysis. However, you are a valued, respected employee who used accepted mathematical methods and have obviously given thought to what you wrote. So what they probably will do is give that one-page memo with the correlation results to one of their high-level analysts, a MBA-type. They will check the analysis, maybe add more variables and then decide whether the idea merits further consideration. If they agree and they change the policy then you will get satisfaction, recognition and possibly more. Companies are listening to lower level employees more these days; it's part of several modern management paradigms. Also, when mathematical analysis supports an argument, it is more convincing.

All kinds of people can do quick correlations and regression analysis on company data. Administrative assistants, lower-level supervisors, secretaries, support personnel, engineers and production line workers all might have something to contribute. They don't have to be managers or MBAs. They might be members of process reengineering teams or total quality management teams that are studying how to improve business practices. Mathematical data analysis is an accepted business practice and everyone can do it, to some degree. Data is everywhere. Do simple data analysis on the data sets closest to your personal business area. It won't take long and it might give you insight into your specific area. Maybe what you gain is better understanding of why your company works the way it does. That alone makes you a better employee. Isn't that worth some effort?

Company Policy Analysis With Cause-Effect

Remember we are not trying to model the entire company operation. We are looking for ways for you to make a difference, you the expert in a small area of your companies' business. We are not trying to make you an instant MBA; we are seeking a way for you to tie together what you do with higher goals and problems. Therefore, we are *restricting our modeling to how your business area effects company goals*. Clearly our models risk being incomplete since we may not be considering all the company data. Our plan is to enhance your unique business-area knowledge with some data analysis to come up with suggestions or ideas that will be studied in more detail. In symbolic form when these two things are added together the sum is suggestions for business policy improvement.

	Your business-area knowledge
+	*Quick data analysis with your business area data sets*
	Suggestions for business policy improvements

The question is: "What can be done at your level, what changes can be made to policy, what regulations or programs can be changed to improve the companies' performance toward one of its goals?" This statement includes both cause-effect and What-If modeling issues. Cause-effect logically comes first, because before you can do any math modeling you must know the independent and dependant variables and that implies a cause-effect relationship between these variables. For cause-effect, correlation is one simple effective way to start. The two data sets should be chosen from a list of potential candidates of the following two types.

Data Sets for Policy Recommendations	
Company goal data (**effect**)	Data representing a company goal or problem.
Your business area data (**cause**)	Data representing a parameter that can be controlled or changed by a change in policy, although the change might require other changes. This data represents parameters, programs, regulations and business practices in your area. Therefore, you are well qualified to recommend changes in these parameters. *Your business-area knowledge* becomes *part* of the *model*.

Usually you will have several data sets in each category and will want to run correlations between each pair from the two categories. Your next step is to consider those pairs of data that exhibit strong correlations. Go through the following type of analysis. Is data set A an effect of data set B? If, no—move on to the next pair. If, unsure—hold this relationship until you have finished the rest of the data pairs. If, yes—consider doing a linear regression between A and B with B as the independent variable. Check the results against your experience, does it make business sense? What range of values of B will this relationship hold for? If the relationship is not linear perhaps add in other cause data sets and do a multiple regression. Data set B might be one of several contributing factors to data set A and simple linear regression won't capture the relationship, multiple regression might be necessary. Or perhaps data set B is correlated to a third data set C which is an actual cause of B. After doing this, you will have a better understanding of what the correlations mean in terms of cause and effect.

At this point you are on your own. Simple math analysis has given you good leads and probably new understandings. Even more, your analysis has a solid foundation since they widely accept correlation and regression analysis. Now you need to decide by yourself what to do next. You are the expert in this business area so it's your call as to what these correlations mean. Math will take you just so far—at some point you have to actually think!

A Practical Example

General directions and rules are nice yet it is good to see a concrete example of this method. Consider a hypothetical but a representative database from a large company. Each row represents data about an individual employee. The column value represents the employees' salary, their yearly evaluation (scale 1 low to 5 high), their education level (1 = HS, 2 = college degree, 3 = graduate degree), their duration of employment, a score they attained on a test given to all applicants, and finally sex (M=1, F=2).

Salary	Evaluation	Education	Duration	Test	Sex
25,000	3	1	3	70	2
29,000	3	1	4	74	1
47,000	4	3	10	81	1
65,000	5	3	7	93	1
32,000	2	2	10	85	1
36,000	4	2	7	84	2
47,000	4	2	6	82	2
28,000	2	2	6	77	2
54,000	4	3	2	86	2
51,000	3	1	14	78	1

Let's look at some data pairs and their correlations and see what that might say about the company's hiring, evaluation and promotion policies. First consider salary. Ideally, salary represents worth to the company. Usually high salaries correspond to managerial, technical or highly skilled positions. Here correlation of salary to performance is .76 which shows that the company does tie salary to performance. Salary and duration of employment are weakly correlated at .24 which shows that your company does not simply reward people for longevity and not performance. Another company-level data set is performance. The company wants to hire and retain workers that perform well. Of course you can't always tell how an employee will turn out but your company tries to hire college graduates for many positions. Also, they do use a company-wide test to screen applicants. How do these two hiring policies work? Are they correlated to performance? Both have a moderate correlation .58 for the test and .56 for education. So to some extent these hiring policies are attaining the desired results. How about discrimination, are they treating the sexes fairly? The correlation of sex to salary is -.27 which means men are paid more. However, that could depend on job type (even with the same evaluations engineers and technical types would probably have higher salaries to remain competitive with the industry). Most companies have similar correlations. How about performance--are women evaluated equally with men? The company evaluation scale is the same for managers to secretaries so each job category should ideally have the same average evaluation score. Your company seems aggressive in trying to keep it equal between the sexes in this area and the numbers bear this out: the correlation between sex and performance is 0 for this small data set. A last correlation is between duration and evaluation which is -.15 which shows a slight tendency for poorer performers to stay and good ones to leave. These are the types of insights that you can get into your company's policies by simple correlation analysis.

Cost-Benefit Analysis

Cost-benefit studies are a mainstay of American business. It works like this. A company considers changing a business practice; often this will cost extra money. The company hopes to get extra benefit (usually profit) from this business practice change. Tradeoff is the key. How much extra benefit will the company get from that extra cost? That is the question. Governmental organizations are deeply into cost-benefit analyses, too. Government programs are always in need of changing or fine tuning. Forecasting the effects of these changes is essential. Mathematics plays a big role in cost-benefit studies. Here the mathematical road starts getting steeper. We are approaching MBA-land. Still, you can do a lot with some simple correlation and regression analysis.

Once you see what is needed you will think of examples and possible applications from your own business area and its data sets.

Examples of cost-benefit analyses surround us. You are considering spending an additional week per year per employee on training. Will the additional profit from a more highly trained workforce make up for the additional cost of the training and the loss of a week of work time? This looks clear enough, notice that it fits in with our theme of lower level factors effecting company-level policy. The problem is measurement and especially measurement in terms of dollars since we need to be able to measure cost and benefit in comparable units—dollars in this case. We can't directly compare 12 apples to 10 oranges; however, if apples sell at $2.50 a dozen and oranges at $4.00 then we know that the 10 oranges cost more.

For our purposes a simple cost-benefit study needs to be done in two stages: 1) determine the conversion formulas from the data sets to dollars and 2) determine a functional relationship between the goal data set and one or more cause data sets. Caution: there is *no* guarantee that this is possible in your case—quite often you can't find a simple conversion to dollars. The case of an extra week of training is a good example: the cost of the training is easy to convert to dollars, simply add the cost of the training to the lost production value. The benefit of the training is nearly impossible to convert to dollars. To determine the value of training you soon get lost in a quagmire of unknown functions and relationships. How much does the training improve the individual worker's ability to do their job? (Where do you get such data and how would you measure 'improve' anyway?) How much company profit would this level of employee improvement lead to? (Same question as above.)

Assuming that you can find a conversion to a common unit then the second step can be done in several ways. You could form a table or a regression relationship between the cause variable and the goal or effect variable. Or you could form a more complex mathematical functional relationship if there is more than one cause variable. Let's see how this works in practice

Katie's company has several product lines and its employs many semiskilled workers to manufacture and assemble products. These are fairly good jobs but are dead end. There isn't any advancement path inside or outside the company. The jobs require some on-the-job training and then the employees are at full production capacity. There are two different divisions A and B within the company at different locations; both divisions employ the same type of worker and build the same product. They organized division B a few years ago. In the spirit of helping employees with their careers, division B started a program with a local community college. This program let these workers take college classes at company expense, in company facilities, on company time and paid the workers' salaries, too. These classes wouldn't be work related. The company had already trained the workers and they needed no advanced skills. It was just a goodwill gesture by a community-minded manager who had enough clout to start it over all the objections. And there were objections, since this program seemed to be a complete giveaway. The courses had nothing to do with their jobs. Employees were simply taking classes in computers, medical assistant training, personnel management or business then leaving the company to take higher paying jobs elsewhere. A complete giveaway. The company incurs a large cost and gets no benefit at all. Right? Or was there something that they were overlooking?

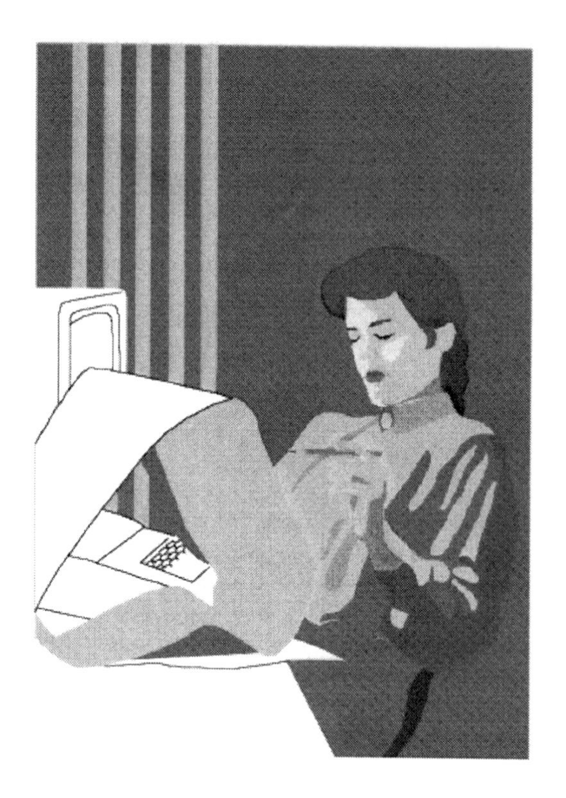

Katie works as a personnel assistant in a large corporation.
Management wants to cancel a training program for semiskilled workers.
How can she use data analysis to convince management to keep the program?

Katie: "Sometimes I just want to scream. Personnel policies can just get totally lost in a large corporation. Really the types of people, their abilities, their surroundings, and their training all contribute greatly to the success or failure of corporations. I just wish management knew that as well as I do. Policies do affect how the company will do but my management is always looking forward to next quarter's new products or industry-wide trends or the stock market. I am just an assistant. How can I be heard? You need a college degree in business administration to be considered for management around here. I'll never get that—absolutely no interest, but I love my job and I have some good ideas. The problem is getting peoples' attention. A solid trend analysis would get their attention. Can I really do trend analysis on personnel data?"

Recently, the influential, community-minded manager, who started the program, left the company. Now upper management wants to kill the program. Katie always thought that it was a good program. She knew several young single mothers that took college courses, got certificates and then got a better paying job with a future in another company. But Katie knows that profit is the bottom line and from all appearances this program doesn't add a cent to profit, but costs a lot of money. Casting around for ideas about how to save the program, Katie examines some other company goals besides profit. She considers employee happiness, an educated workforce and low employee turnover. Katie knows that after one of this type of employees has been on the job a few months that they are as good as they are going to get. More education doesn't help on the job. Also, surveys have shown no difference in employee attitude between those taking courses and those not.

All discouraging, but Katie decides to look at the turnover data. With a spreadsheet she looks at the average duration of employment in divisions A (which doesn't offer the program) and B (which has the program). More discouragement: people stay longer in division A. But that is illogical, the program is an incentive to stay. Why don't people stay longer in division B? More data analysis shows why. Division A is older and its workforce is older. It has not had the turnover associated with newer, younger workers. This leads her to an idea; she narrows the data down to the group of employees hired since the program started in division B. She checks the average duration of workers hired since the program started. She finds that these workers taking college classes stay longer than those that didn't. It is this data set that represents the effect of the program on the company goal of low turnover. Therefore, it is the goal/effect data set. The cause-data set is the number of courses taken by the employees hired since the program started. Next she runs a correlation between the length of time spent at the company and the number of college courses taken. The two data sets are highly correlated. This makes sense. So the program does enhance one company goal, lower company turnover. People stay longer if they are enrolled in the college program. That is something to point out to upper management and the correlation calculation supports it. At this point, she could make a suggestion that the program lowers turnover among new hires citing the correlation results as evidence. Still, management could dismiss that. Management might decide that the factors can be correlated and the program still could be a money loser.

Next Katie wonders could this extra duration of employment due to the college program saves the company money since they don't have to hire and train new workers so often. She asks herself if the goal data set: duration of employment (equivalent to turnover rate) can be given a dollar value. She already knows the cost of the program: $2,000 per course per employee (total of tuition, books, salary for time spent in class, lost production and cost for company facilities). She also knows the cost of hiring and training a new employee, $19,000. Next she takes that pair of correlated data sets and fits a linear regression model to the goal/effect data using the cause data as the independent variable. The result comes out as (units are months) $f(n) = 5n + 18$ or in words: *expected length of employment* = *5 * number_of_classes* + *18*. This says that a new hire who doesn't take any college course averages 18 months on the job. Also it says that a new hire that takes 2 courses during her stay in the company averages 28 months on the job. Clearly, people that take classes turnover slower and therefore the company saves more advertising, hiring and training expenses. However, is the savings worth the cost?

The tradeoff is this: cost = $2,000 per course per worker; benefit = less turnover per worker. The term "less turnover" needs to be quantified and converted to dollars so that it can be compared to the cost. A key point is that less turnover or longer length of employment means the company saves money by hiring and training fewer people over the long run. There are several ways of looking at this. For example 28 employees who each stay 18 months (the group that didn't participate in the program) would have a total of 504 work-months. This 504 work-months is the same as 18 employees whom each took 2 courses and stayed 28 months apiece. The company saves the hiring and training costs for 10 employees or $190,000—that is the benefit. The cost is 2 courses for each of 18 employees at $2,000 per course or $72,000. Therefore, the savings associated with the college course program would be about $118,000 for this amount of time, 504 work-months or about $7,000 per employee.

Now this proves it! Upper management has been thinking that the classes simply waste money because they can't see the connection between the training and their personnel costs. Katie has used her in-depth knowledge of the personnel data and some simple data analysis techniques. She has shown that the "worthless, wasteful" training program actually keeps people working longer at the company just to take the training for their future. This "working longer at the company" translates into a savings of costs to find, hire and train new workers. The benefit to her company is more than the cost. So the program is actually a money maker. "Keep the program" she suggests. (P.S. Her company did keep the program. Moreover, expanded it. Katie got a cash award for her suggestion, but no promotion. However, ever since when she voices an opinion in the weekly staff meeting—people listen carefully.)

What happened here? First, Katie knew a lot about personnel, much more than upper management. She knew the turnover statistics. She realized that division A was older, had older personnel whose turnover rate was less than newly hired people. She also knew that the new hire workers weren't using the income as a second family income but instead as their main source. This expert local area knowledge allowed her to reject the superficial conclusion that division A had lower turnover so the college program wasn't helping decrease turnover. She looked deeper and found the correct cause-data set which was the set of employees hired after the program started, not all the employees back into prehistory. Second, she could convert both data sets to comparable numbers: cost and benefit both in terms of dollars. That is seldom easy and not always even possible. Third, she used correlation to find a relationship between low turnover and the number of courses take. That fact in turn led her to try linear regression between these two data sets. After linear regression she had a fact that nobody else in the company knew: a model equation $f(n) = 5n + 18$. It was a killer fact! It allowed her to do the cost-benefit study and convince management to keep the program. Katie's expert local area knowledge plus some simple statistical tools gave her the advantage in decision making as seen in the following table. Notice upper management was acting on general or macro level business judgements (they often do) plus some superficial data analysis. Katie went deeper. Notice too, that the last level is normally MBA land. Earlier, I recommended making suggestions based upon expert knowledge and correlation analysis. Katie could have stopped here and done that (the third level). However, she chose to do the next level, cost-benefit study, by herself. This was great for her and will be for you too, if you can do it. Often the math is too much and you should pass the idea along after doing correlation. Don't worry, good ideas will still make it through.

Levels of Analysis

Analysis Type	Recommended Action
Macro level business judgement The program costs the company. The education is useless to the company. Conclusion: The program loses money.	Cancel the program.
Macro level business judgement plus superficial data analysis The program costs the company. The education is useless to the company. The program loses money. plus Average turnover in division A (no program) less than in B (has program)	Cancel the program.
Correlation analysis New employees who take the courses stay on	Situation not clear

the job longer than those who don't. The program is not useless; but more analysis is needed to see if it is profitable.

Cost Benefit Analysis with regression model Employees who take classes stay an extra 5 months per class. The savings of hiring and training costs is more than the cost of the courses.	Keep the program.

Checklist

The chapter is not meant to make you an instant MBA. Nor does it give you a recipe for far-ranging data analyses. It focuses on a specific situation, which is capsulated as follows.

Specific Business Situation

You are an expert in the policies, processes and data in a specific area of company business (e.g., personnel, production, training, engineering, etc.) Can you use *your expertise plus data analysis* techniques to recommend changes in *your* area that will improve company goals?

Data analysis enhances your own expertise; it identifies specific cause-effect relationships, it quantifies parameters that depend on other parameters and it gives a definiteness to your work. Data analysis does *not* replace the need for knowledge or area expertise. The following checklist is a good place for you to start. It isn't all-inclusive; there will be some situations that you need to work out on your own. Still, the checklist is a good place to start for your first time. Remember the goal is a recommendation to your management about how a change in your area can improve overall company performance.

1. **Identify goals, causes and data sets**: Identify a company goal/effect that might be effected by processes in your area of expertise. Identify a collection of data sets of both types, goal/effect and causes.

2. **Cause-effect analysis**: Calculate correlations between pairs of goal/effect and cause data sets. Review and analyze the results using your specific knowledge about the underlying business processes. Are there true cause-effect relationships at work here? If so go on to step 3.

If not then you might have to consider ending your analysis at this point and deciding if it merits sending along as a suggestion. If you can't identify cause-factors in your specialty area, your suggestion might lack a solid foundation for suggesting change.

3. **Modeling**: If you have a functional relationship then you can try modeling the goal/effect as a function of the cause data sets. (Regression or multiple regression is a good place to start. Neural network analysis might be suspect to some upper management who don't understand it.) Ask yourself: what changes would have to be made in company policy or processes to make the cause parameter change? Does your model support your logic? If not, go back and try again.

4. **Cost-Benefit Analysis**: Can you find a conversion from your goal/benefit data set and your cause data set into a common unit, like dollars? If so and if you can model the relationship then you maybe able to do a cost-benefit analysis. Consider the possibility.

5. **Recommendations to Management**: Package together your reasoning and data analysis results. These could be of the form of a simple logic based on your business savvy and simple correlations or it could include math models or a cost-benefit study. Then send this as a memo or e-mail to your boss (or to a knowledgeable peer if you are nervous about making a suggestion to your boss). Or do a quick presentation at a staff meeting.

Chapter 12

Real World Applications

Final Advice For Your Data Problems

Have you seen your own data problem yet? Does any of this apply to you? It should, to be sure we are going to go over common real world application areas. The stock market, sports betting, real estate investing, and collectibles are other areas where real people invest time and money. Regular people don't have math degrees or backgrounds in statistics; yet regular people want to do a better job and make more money by data analysis. Can you develop an incredible software prediction method and get rich quickly? No, probably not. Can you understand your data better, make better decisions and make more money? Yes, that you can do. Let's look briefly at several common situations.

The Stock Market

You might think that predicting the stock market is an exercise in economics theory: that underlying economic forces are what drives stock prices; so what you have to do is the model the near-future economy. Maybe there is some truth to that; there is an entire school of stock market thought that says that trading on market "fundamentals" is the way to go. However, market psychology plays a larger role in the day-to-day swings of stock prices. Investor confidence ebbs and flows; their evaluation of stock prices changes too. This means hundreds of thousands of investors are trying to outguess each other about the market trend. This means thousands of computers running everyday trying to outsmart each other in market trend prediction. This means your trend prediction is not modeling just the economic trends but also it is trying to outguess what thousands of other programs are doing. One major point of reference: over the past ten years most professional fund managers (who usually have enormous computing capabilities) have not done as well as the standard market index, the S&P 500. Random guessing can often achieve same return as the S&P 500. This has been done often in the past by throwing darts at the market page of newspapers to choose the next stock purchase. This point is made to instill one thought—beating the stock market with computer analysis is one tough proposition. You have been warned.

Now if you still want to try, you need a method and some software. For general advice, I can do no better than recommend any of several books on technical analysis of the stock market. There are more data analysis, trend prediction, pattern recognition and classification methods around than I can possibly cover. For computer software there are two major types of data analysis software: general software (like spreadsheets, statistical programs and neural networks) and specialized financial investment software (aimed at stock market data, with stock market tutorials and graphic presentation modules that have standard stock market graphics already programmed in). They often design these specialized financial investment software packages around statistics, neural network analysis or expert systems. For the beginner, this is probably easier than buying a general neural network or statistics package. Because then you must obtain stock market data and format it in a proper way to input to the general package and format the output also. If you are good with computers, don't shy away from good general data analysis packages; stock market data comes in standard formats that most modern data analysis software can easily handle. This is a tough problem and there are more data than you can possibly analyze. However, success means big profits, so this area has always attracted data analysts. One good thing about investing in stocks is that it is a positive sum game. This means that overall stocks increase in value, so people are profiting from a value increase not profiting necessarily from other peoples' losses—like in gambling.

Love and Marriage

The social sciences are awash in statistics and trend prediction. Most large universities have hundreds of researchers compiling data and predicting social trends. One of the most analyzed areas is marriage. Marriage might seem a strange thing to subject to cold, impartial trend analysis. Yet many local, state and federal agencies track marriage and divorce statistics routinely. As a nation and a people we need to know what direction we are headed in and where we came from. So therefore, there are many statistics, reports and books about marriage trends, age-at-first-marriage statistics, and projections. Over the past two decades two trends have been apparent: first, there have been fewer marriages and second, the average age at marriage has been steadily rising. Sociologists that follow these trends attribute much of it to a stronger women's presence in the workforce. There is less of an economic reason to marry and even less to marry young. Men are not taking it well either, sexism is a factor. Most men seem unwilling to help with their share of housework and child rearing, which a working wife and mother needs. Other reasons involve large scale social changes in the nation, the media, the schools and the populace itself.

In the mid 80s there was a lively debate in the media about work done by some researchers, most notably Neil Bennett (Yale) and David Bloom (Harvard). They published work indicating (among other things) that nearly 1 out of 5 college-educated women born in the 1950s would never marry. Also, they wrote that their work showed that 1 of 8 women born in the mid-1950s would never marry compared to only 1 of 25 of the preceding generation. The media jumped on this, expanded its meaning. For months there were columns making wild claims like: a single, college-educated, 40-year-old woman had a better chance of being killed in a terrorist attack than ever marrying.

Bennett and Bloom got battered from both sides: feminists and conservatives. Researchers from the Census Bureau jumped in with a study of their own that claimed to refute this study. Bennett and Bloom fired back. They said that this Census Bureau study based on earlier data didn't project trends as well as their study which was based on Census Bureau data from the mid 1980s. The battle raged on. It still hasn't entirely stopped. The following graph is a conglomerate graph that shows the distribution of ages that women marry at. Men have a similarly shaped graph with a slight time-shift to the right since they tend to marry about 2 years later on the average.

Trend analysis was the methodology for these studies. The entire debate was based upon different methodologies for trend analysis and different assumptions about function forms. Data is plentiful about marriages, ages and divorces; but getting a representative sample isn't always easy. Also, different surveys give different results because the underlying population sampled differed. It's not easy. That is why trained statisticians abound in this area. The popular writing done in this area neatly and concisely states facts like "1 in 5 college-educated women born in the 1950s will never marry." The facts are a bit murkier since the research papers that these facts are derived from are not easy reading. Still, trend analysis is vitally important in these sociological areas too.

Many mathematical curves can be fit to this data. Most of our examples have been simple polynomials: linear, quadratic or perhaps with a sine function added. Other common math functions such as logarithms, exponential, trig and more complex functions are used too. Different functions give different results. Sometimes the differences are trivial; sometimes they are not. For example on this graph a key point is what percentage of all women (men) will eventually fall under the curve (eventually marry). People can and do marry for the first time in their forties and fifties, no one denies that. The point of contention is whether that number will be sufficient to keep the overall marriage rate above the 90% historical figure. The debate continues.

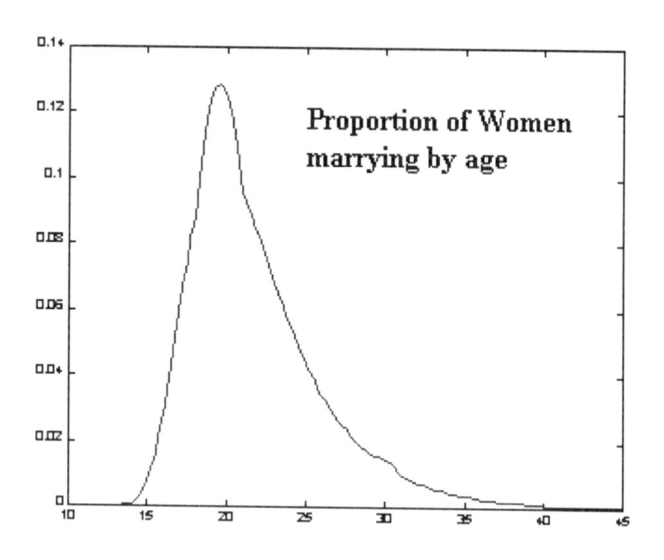

Sports Betting

Betting on sports is big business, billions each year. Much of it is illegally wagered, as most states don't allow this type of gambling. Some states have state controlled lotteries, some have low-bet-limit casinos or river boats, several have racetracks, however, sports betting is generally illegal. Recently, offshore Internet sites have offered on-line betting but betting with them is also considered illegal by the Justice Department. Las Vegas has everything, of course. Data analysis is completely irrelevant in standard casino games like roulette, craps, slots or lotteries. These games are simply random chance with a percentage for the house or state. You pay your money and usually you lose part of your money. Hopefully you have a good time while you are doing it. Sports betting—football, basketball, and baseball—is a different story.

The key point in sports betting is that you are not really betting that one team beats the other team; you are betting that one team beats another team by a certain number of points. There is a world of difference. You aren't trying to pick favorites to win the game outright, which is easy to do more than half the time. Teams can go unbeaten all season and their fans who bet on them each week can lose money! A cellar team with a 4-12 record can make its betting supporters a bundle because they stayed close while losing.

Sports betting in its simplest form is a bet on (say) a football game where (say) the Cowboys play the Cardinals. Since the game is not an even match bookmakers assign a point spread to the game and they might list it as: Cowboys

(-7) vs. Cardinals. It's like handicapping in golf or bowling. The bet is decided by whether the Cowboys beat the Cardinals by more or less than 7 points. (Of course they decide the real game simply by the score.) People set this 7-point line—it is not a mathematical truth like the odds at roulette or craps or lotteries. No one knows for sure that if the Cowboys play the Cardinals play 100 games that the average difference in score will be 7 points more for the Cowboys. The bookmakers could be wrong. Sports bettors are an opinionated group; they bet on games for a variety of reasons: maybe one team is the home team or they simply don't like one of the teams in a game and bet against them. Most bettors feel that they can call the winner in a large percentage of games and that the line is just an additional hurdle that they have to jump. Nevertheless, a good line makes any matchup, even mismatches, a fairly even bet. That is why the sport betting business thrives: the bettors win and lose about evenly and the house gets its steady percentage.

Back to the Cowboys and Cardinals: if the actual expected difference in abilities of the two teams is actually a 10-point spread for the Cowboys then the Cowboys' bettors have a large statistical edge. If you could find games every week with spread mismatches like this, you could get rich. However, Vegas has the experts, Vegas has the data and Vegas has the computers and there are large betting groups out there with their own experts, data and computers. Can you really beat the Vegas line? Tough question, a really tough question and in some ways it is similar to the stock market prediction question.

First, the percentage: usually you have to bet $11 to win $10. The difference is the "Vig" or house take. This figures out that you have to win about 52.4% of the time just to break even. If you handicapped every NFL regular season game, you would need to call about 126 right out of 240 games to break even. That may not seem like much, but it is. Many people feel that their own special understanding of football will more than compensate for this. You may be one of these. In this regard, check out this one point: many large newspapers have weekly columns that pick next Sunday's games; these writers are football experts, they are knowledgeable about the current injury status, predicted weather conditions, history between the two teams, team trends and other information. Check out the record of your favorite columnist for the past few years. He probably didn't beat these odds much if at all. Nor do professional handicappers who make predictions in betting newspapers. It's tough.

Now the question is: "will extensive computer data analysis help?" This is not clear—but I doubt it. Extensive computer data analysis went into making the line in the first place. To beat it, you must consistently find situations with your computer like where a 7-point favorite is actually 9 or 10 points better than the other team. That is going to be extremely difficult. In addition, the type of computer analysis is an issue. Recall that we talked much earlier about models developed from first principles, from first principles plus data and from data alone. Modeling from first principles generally is more accurate. Yet, often in the business place the underlying reality is so hard to figure out that you usually have to model with data alone. The stock market maybe the same situation. However, that is not necessarily so for sports betting. There are excellent sports simulations out there; if you don't believe me, try out some video games like NFL GameDay or some of John Madden's pro football simulations. These programs are the ones that they can package for PCs and sell to the public. You can believe that there are some super simulations of football and other sports that help set betting lines. You might be a computer expert and a savvy football analyst; you might feel that you are far ahead of the average bettor in these respects (you are probably right). Still, those two talents may not be enough.

The key point: you aren't betting against ignorant masses that place bets as much as you are betting against highly skilled computer specialists and football experts that set the line. You might think that the guy that you met at the corner bar is a complete fool to bet on Green Bay at -11. You might have listened to his analysis and dismissed him as totally ignorant of football. This might make you feel better about betting against Green Bay—after all with fools like this betting against you, doesn't that give you an advantage? No, it doesn't. If the bookmakers set the line properly at -11, then fools and wise men alike are facing the same odds. If all the fools in the world bet a game one way and all the smart people bet it the other way and if the line was set accurately then it is an even bet. Your only chance is when the line is set differently from the true odds (whatever they are). Bookmakers don't want to risk too much on single games so if too many people bet one way they will move the line. Perhaps the move will be away from the true odds. Therein lies your chance.

All in all, sports betting is not a money making proposition and it's illegal in most states anyway. My final advice: watch sports on TV while your broker invests your money in the stock market.

Real Estate

Real estate is a major source of investments in this country. Real estate is different from the stock market; there aren't these day-to-day fluctuations in prices and seemingly random trends. Real estate is usually more predictable and often safer. In addition, you can touch something tangible. Real estate transactions cost you a fee like a brokerage fee in stocks or a bookies' vig in betting. However, the overall trend in real estate prices is up, so like the stock market many people profit, it's a positive sum game.

Can you use data analysis, neural networks or regression to do real estate trend predictions and pricing? Yes, you probably can. For example: you have modeled the private home market in your area and have found a powerful new demand for two-car garages, because there is an influx of two-income families with kids and vans. Most older homes don't have this feature. This is just a temporary glitch in the market; new home builders are building mostly two-car garages and they will eventually satisfy demand. Your program fits all the pricing data and finds some homes on the market that probably could sell for $40,000 more than their current asking price, if they had those attached two-car garages. Maybe you could build the garages for less and resell the house. This is "niche" investing; you probably don't have many competitors in your local area. Certainly, you won't have thousands of market analysts with supercomputers or a million Cowboys' fans to compete with.

The issues of computers vs. humans and how prices are set come into play here. For comparison: in both the stock market and sports betting you are definitely competing with supercomputers who are helping set stock market asking or selling prices or the line (which is like price) of a game. In real estate there are some legal issues; assessments of real estate often must be done by set methods and licensed Realtors. They have taken courses, passed tests, are experienced and are familiar with the local economy and the makeup of the local population. However, data analysis methods like neural networks or regression methods are not necessarily doing assessments or pricing. (There are many caveats since real estate laws vary widely as do real estate company methods.) Also, like Katie in the large corporation example, you might have a lot of pertinent knowledge about the local area, because you have lived there for a long time. You might have a lot of experience with the economy, the development issues and other things that give you an investment edge. Therefore it is entirely possible that you could design a neural network or regression method that better predicts pricing or demand for some types of real estate in your local area. "Better predicts" means an edge for you that could translate to profits.

Collectibles

Millions of people collect things: baseball trading cards, coins, antiques, dolls and stamps to name a few. These people spend large sums on these collectibles. Over time prices change. A rare Roman coin that was worth $27 in 1989 might be worth $50 today; this could be a source of profit. The question: could we use mathematical prediction techniques like regression or neural networks to predict the change in price of collectibles?

Consider coin collecting (the others have similar characteristics). There are thousands of types of coins out there. In just this country, in just this century, we have minted hundreds of different types of coins: 1917 pennies, 1926 pennies, 1947 nickels, 1949 dimes, hundreds of denomination-year combinations. Check coin catalogues, these all have different values. Now go back two centuries or consider all foreign coins or all ancient coins. Thousands of variations all with different values; values that can change depending on availability and tastes. Next consider the available data and the prediction problem. You have available a time history of the prices for each type and condition of coin. Can you predict the future price trend accurately enough to speculate? Possibly, but you should know a lot about the coin collecting business: rarity, coin condition, demand, customer type and especially trends. Like the stock market, past time histories are not sufficient alone to predict future prices. Time histories are an important source of data but these other parameters need to be included in a neural network or regression program. Also, with literally thousands of different coins as potential patterns/outputs for a neural network you must drastically restrict the number of coin prices you try to predict or your neural network will never finish. With these cautions in mind, an insightful, clever coin collector might use math modeling techniques profitably. However, as with the stock market, sports betting, and even real estate transactions—you pay to play. If you deal in coins through retailers then they attach a profit margin that is your cost in coin speculation. Other collectibles are similar in nature. Also as with Katie, that large corporation administrative assistant, who had a great deal of knowledge of her local area, your expert knowledge will be of enormous help is setting up any data analysis method. If you don't have this knowledge, you might reconsider. My advice: collectibles are something you love; if you don't enjoy it as a hobby first, I'd recommend trying something else for sheer investment potential. It's those details like rarity, demand and condition that you need to include in any prediction and who knows these better than someone who is intimately knowledgeable—the avid collector.

Software You Can Use

In the course of this book we have mentioned several types of commercial software: spreadsheets, database systems, neural networks, statistics and other more specialized types like forecasters, expert systems, solvers and more. First, a general statement: there is a lot of good user friendly software out there; if you are a beginner most of it has far more capabilities then you need. With that in mind we will quickly wrap up by going through the general categories of software and discuss some specific packages available.

Major office suites like Microsoft, Corel WordPerfect and Lotus offer excellent spreadsheet programs (Excel, Quattro Pro, Lotus 1-2-3) as part of a suite of software that includes word processors, database systems, and presentation and graphics software. Add-on software like solvers are usually available for these spreadsheets. In addition, neural network, data mining and forecaster software usually gives easy interfaces to these spreadsheets, because these spreadsheets are so popular and widespread. For this reason spreadsheet programs are often a starting point for predictive data analysis.

The major suites also usually include a database program as part of the suite. These major vendors offer Access (Microsoft), Paradox (Corel WordPerfect) and Approach (Lotus). Any of these will give you a good start with database analysis. If you don't want to spend that much to start with, keep in mind that often your PC will come bundled with quite usable spreadsheet and database software; you might just use that until you become more expert and decide exactly what advanced capabilities you need.

They include some statistical and mathematical capabilities in spreadsheet programs. If you decide you need more, than two excellent mathematics packages with many add-ons are MathCad, MatLab and Mathematica. Pricing is difficult to assess; because you have options of buying add-on packages at low prices or buying a lot of capability for more money. If you are a good programmer in C or Fortran or Pascal than you might consider a book *Numerical Recipes in C* (versions in Fortran and Pascal also) which has hundreds of mathematical routines already coded in C and many examples and advice about how to run them. This approach would be a little more advanced than a software package.

Neural networks come in a wide variety. An important point about commercial neural network software: neural networks are not as widely accepted as spreadsheets or statistics. The name sounds too advanced for many buyers (we know that this is not true, neural networks can be simpler than some spreadsheet analysis or statistics, but that is the perception). Because of this the market for neural networks is just starting and it is more of a niche market. Prices might be higher for a neural network software package than you might be used to paying for other PC software. However, several vendors offer a range of products varying from a hundred to several thousands of dollars. The examples in the book were run on two neural networks that I use: Brainmaker by California Scientific Software and ModelQuest Miner by Abtech. Both were the basic models and (at the time) I got each for $150 or less. They are different: Brainmaker is more of a standard neural network, working with backpropogation and sigmoid functions; ModelQuest Miner is something of a cross between neural networks and statistical regression. Since they represent different approaches to neural networks, I like to run both of them against my data and compare results. If you are just starting and don't want to spend much on software you might try a book by Timothy Masters, "Practical Neural Network Recipes in C++." It includes a neural network program that you can run, along with sample files and advice. Also, there are literally dozens of free or shareware programs available at AI sites on the Internet.

Expert systems are also available. A good place to look for AI software is "PC AI" a monthly magazine devoted to Artificial Intelligence software and techniques; they often list vendors, products, phone contacts and pricing in tables at the back of the magazine. They discuss both neural networks and expert systems a lot in this forum. Also, for financial or stock market predictions there are several software products including neural networks, expert systems, combinations of the two, forecasters and other hybrid systems advertised in stock market analysis journals.

There is a lot of good software out there. One thing a first-time buyer might want to avoid is overbuying. If you are just trying out a new approach, it doesn't make sense to spend thousands of dollars of your own money to experiment; you can probably get a good system for much less and then if you decide that your own problem requires much more capability then you can buy the expensive model.

Bibliography

Dewdney, A. K. *200% of Nothing* (New York: Wiley, 1993) Examples of abuse of statistics, math and logic.

Epp, Susanna *Discrete Mathematics with Applications* (Belmont, California: Wadsworth, 1990) The only actual textbook in this list. The first two chapters have a great presentation of basic symbolic logic. I have taught from it for years.

Freedman, David H. *Brainmakers* (New York: Touchstone, 1995) A general nonfiction book about AI. Interesting reading.

Huff, Darrell, with Irving Geis *How to Lie with Statistics* (New York: Norton, 1954) A classic book. Still readable and useful.

Lawrence, Jeannette *Introduction to Neural Networks* (Nevada City: California Scientific Software Press, 1988) An excellent introduction to neural networks, how they work and how to apply them to real problems. If you want to learn more about neural networks, than read this.

Maturi, Richard J. *Diving the Dow* (Chicago: Probus, 1993) If you want to invest in the stock market using trend prediction software this is a good short book on stock market indicators and trends.

Paulos, John Allen *Innumeracy* (New York: Vintage, 1990) Lots of great examples of abuse of math, logic and statistics. Paulos may be the best of modern day popular math writers.

Paulos, John Allen *A Mathematician Reads the Newspaper* (New York: Basic Books, 1995) More great examples of abuse of math, logic and statistics.

Rucker, Rudy *Mind Tools* (Boston: Houghton Mifflin, 1987) Math logic for the layman is part of this book.

Seiter, Charles *Everyday Math for Dummies* (Foster City, CA: IDG, 1995) Another fine "Dummies" book. Some tips about algebra, geometry, trend analysis and mathematics for different business concerns.

Stein, Sherman K. *Strength in Numbers* (New York: Wiley, 1996) A couple of chapters (9 and 10) discuss the levels of math needed for different occupations. This kind of information is hard to find outside of rather dry Bureau of Labor reports.

Tobias, Sheila *Overcoming Math Anxiety* (New York: Norton, 1994) The book that made the phrase 'math anxiety' a household expression. Discusses both what you can do personally and the politics behind math education. If you have math anxiety, start with this great book.

Vos Savant, Marilyn *The Power of Logical Thinking* (New York: St. Martin's Press, 1996) Logical reasoning and abuse of statistics are brilliantly discussed in the book by the famous columnist.

Index

A

B

C

D

E

F

G

H

I

K

L

M

N

O

P

Q

R

S

T

U

V

W

Author Biography

William May has a Ph.D. in Mathematics from the University of Nebraska and has been involved in engineering for the government and teaching mathematics and computer science at several universities. Previously, he wrote a book *Edges of Reality* about ultimate limits on computers, artificial intelligence, astronomy and human thought. Dr. May has been an Adjunct Professor at George Washington University, George Mason University and for the past 12 years at Virginia Tech. He has worked for several government agencies and engineering companies including the Central Intelligence Agency and the Navy Department. Bill and his wife Rhonda reside in northern Virginia.